Describing Early America

Describing Early America

Describing Early America

Bartram, Jefferson, Crèvecoeur, and the Influence of Natural History

Pamela Regis

University of Pennsylvania Press · Philadelphia

Originally published by Northern Illinois University Press

First paperback edition 1999

Printed in the United States of America on acid-free paper

10 9 8 7 6 5 4 3 2 1

Design by Julia Fauci

Published by
University of Pennsylvania Press
Philadelphia, Pennsylvania 19104-4011

Library of Congress Cataloging-in-Publication Data

Regis, Pamela.
Describing Early America : Bartram, Jefferson, Crèvecoeur, and the influence of natural history / Pamela Regis.
p. cm.
Originally published: DeKalb : Northern Illinois University Press, 1992.
Includes bibliographical references and index.
ISBN 0-8122-1686-5 (acid-free paper)
1. American prose literature—Colonial period, ca. 1600-1775—History and criticism. 2. American prose literature—Revolution, 1775-1783—History and criticism. 3. Natural history—United States—Historiography. 4. United States—Intellectual life—18th century. 5. United States—Description and travel. 6. Nature in literature. 7. Rhetoric—History.
I. Title
PS367.R44 1999 98-47416
508.73—dc21 CIP

Frontispiece and title page illustrations: Koohassen, Warrior of the Oneida Nation, courtesy of the Johns Hopkins University Milton S. Eisenhower Library, Special Collections; and *Anona Pygmea*, drawing by William Bartram, for *Travels* (1791).

FOR MY PARENTS

· *Francis and Ruby Thompson* ·

Contents

Acknowledgments

Professors M. Delmar Palmer, Hugh Kenner, and Larzer Ziff provided support and advice for this project. Carol Quinn, Cheri Smith and Mark Collier of Western Maryland College's Hoover Library zealously tracked down much-needed volumes. Pat Holford helped keep me organized during the initial drafting of this manuscript. My husband, Ed Regis, lent thoughtful attention to several years' talk about plants. His suggestions improved this work in many instances.

I thank them all.

Prologue

Recovering a Lost Paradigm

Critics have read William Bartram's *Travels,* Thomas Jefferson's *Notes on the State of Virginia,* and J. Hector St. John de Crèvecoeur's *Letters from an American Farmer* as works of belles lettres: as sublime and picturesque travel narratives or as novels. They are not. They are works of science.

The science of natural history reached its ascendancy during the mid-eighteenth century. Contemporaries of Bartram, Jefferson, and Crèvecoeur understood natural history to mean a broad area of scientific inquiry circumscribing the present-day disciplines of meteorology, geology, botany, zoology, and ethnology. Natural historians took for their subject matter all of what they called the Creation. Any object within the natural order was a proper subject of natural historical inquiry; only man-made objects lay outside its scope. What we today view as different sciences with different methodologies, the eighteenth-century thinker saw as a single science with a single methodology. Recovering this lost paradigm for investigating natural productions will allow us to understand *Travels, Notes,* and *Letters* in a new way, one that is closer to the way in which their authors originally conceived, organized, and drafted them. The rhetoric of each of these three texts—from its organization to its sentences and diction—is governed by the way natural historical investigators did science.

To recapture this lost paradigm and the rhetoric that it created, I establish the importance of natural history in revolutionary America, both in the broader scientific life of the country and in the lives, thoughts, and texts of Bartram, Jefferson, and Crèvecoeur. I read Bartram's *Travels,* Jefferson's *Notes on the State of Virginia*, and Crèvecoeur's *Letters from an American Farmer* based on the tenets of natural history used in their construction. I also locate these works within a larger group of texts, the literature of place. The works of Major Robert Rogers, William Smith, William Stork, James Adair, and Jonathan Carver constitute a loosely defined genre, a tradition of writing about the land, in which *Travels, Notes,* and *Letters* belong. As explorers of North America, these men observed and recorded basic information about the natural productions in the territories they visited: the rivers, soils, plants, animals, and inhabitants.

Of the various topics that composed natural history, botany held a special place. Natural historians concentrated on the study of plants because they were often of practical value and were easy to collect and display. Old-world collectors vied to outdo one another's stove houses in which they displayed exotic species. It was botany, primarily, that engaged William Bartram, son of botanist John Bartram; Jefferson, the plantation owner and experimental gardener; and Crèvecoeur, American farmer and French author of botanical treatises. They applied the natural historical method that had been developed for botany and, using the resulting rhetoric, constructed descriptions.

This natural historical method employed by Bartram, Jefferson, and Crèvecoeur, a method I explore at length in the chapters that follow, had an unfortunate consequence: it led them to depict human beings as if they were just another type of natural production. "Men," Crèvecoeur said, "are like plants."[1] This epigram states the theme he pursues in large sections of *Letters*. Bartram and Jefferson, less self-consciously, made the same correlation between men and plants. Natural historical discourse did not differentiate between vegetable and human. This had serious consequences for the people most often depicted in these descriptions of the new land: native Americans and blacks. These two great victims of America's founding became, in these descriptions, Other. Of the people they wrote about in *Travels* and *Notes*, Bartram and Jefferson described only those who were most different from themselves—"savages" both native and black. Crèvecoeur applied the natural historical method and rhetoric to European settlers as well, turning them into Others in the process.

Once we understand these texts as works of natural history, we are in a better position to assess their belletristic elements. Critics of America's early texts seek the first American novel, investing Crèvecoeur's *Letters* with novelistic elements. Or they read Bartram's *Travels* and Jefferson's *Notes* as travel literature concerned chiefly with the sublime and the picturesque. Jefferson's descriptions of the Blue Ridge Gap and the Natural Bridge have been read to the exclusion of the other 255 pages of his text. Critics are impatient to see "American literature" blossom forth. By "literature" they have traditionally meant what Stone in his introduction to *Letters* describes as "arrangements of affective images embodied in the traditional forms of poetry, fiction, and drama, and expressing the spirit of place."[2] Although this narrow definition of literature is fading, it has left a residue of assumptions that leads critics to impose belletristic analysis on the texts they read. My purpose is to provide alternative readings of *Travels, Notes,* and *Letters* based on the lost paradigm of the science of natural history so that we may have an accurate account of the intellectual tradition that gave rise to these texts and that determines much of what they mean.

Describing Early America

CHAPTER ONE

Natural History in Context

Courtesy of John Bartram Association.

In 1758, John Bartram sent to Peter Collinson this drawing of the garden that helped secure Bartram's international reputation in botany.

In 1758, on the eve of the revolutionary era, Jonathan Edwards died from complications resulting from a smallpox vaccination. As the last major representative of the Puritan literary tradition in America, Edwards's inward-looking writings reflect the world that mattered to the Puritan—the features of mind and of belief, the relationship between believer and God. Seven years later, Parliament imposed the Stamp Act, and while British literature produced the second wave of the founding works of prose fiction (*Vicar of Wakefield*, 1766; *Tristram Shandy*, 1760–1767; *Sentimental Journey*, 1768; *Humphrey Clinker*, 1771) American letters turned its attention to the creation and definition of the new country. A novel distinctively American in setting and theme would not be published until 1798 (*Wieland*). The prose fiction revolution that was establishing the novel as the preeminent belletristic literary form came late to America. Before novels could be set in America, the country itself needed to be written into existence.

Nonfiction prose was the dominant American literary form from the time that American independence became an issue until the establishment of the federal government. The literature of American self-creation and self-definition in the period from the Stamp Act (1765) to Washington's first inauguration (1789) was largely utilitarian, nonfictional, and nonbelletristic: essays to take the American case to the world's notice ("Appeal to the World," 1769), pamphlets to incite the spirit of independence (*Common Sense*, 1776), a Declaration to effect independence itself (1776). Belles lettres—poetry, the drama, and prose fiction—were secondary to prose that conveyed fact.

In addition to the documents that argued for and established the new government, American factual prose included what Moses Coit Tyler has described as "writings which stand for the delight of man in the visible framework of nature in the New World as hidden in plant and mineral, and animal: descriptions of the American wilderness, narratives of travel among the native peoples, with sketches of their characters and ways."[1] This is a wide range of topics for any single kind of writing to address. But this list of topics—plant, mineral, animal, native peoples (discovered through travel in the wilderness)—comprises an agenda for one more form of self-creation and self-definition to stand alongside the political documents that promoted these two aims. The creators of the new nation had to consider not only the political justification and foundation of the new country but also the description of the territory or ground on which such politics would hold sway. The pamphlets, the Declaration, the Articles of Confederation, and the Constitution itself create and define the new political entity called America. The physical ground is described and, insofar as discovery and description constitute a kind of definition, defined, by a series of texts. Turning away from the interior landscape of the Puritan, writers such as William Smith (*Historical Account of Bouquet's Expedition against the Ohio Indians,* 1765) and Major Robert Rogers (*Concise Account of North America,* 1765) examine the visible world in the unknown interior of the continent, beyond the colonies' westernmost settlements. As more of these texts are written, they form a tradition, a genre. As the genre matures, a second generation of these texts takes on larger, more resonant meanings than the earlier, more fragmentary accounts of place. No longer accounts of first explorations, later texts can encompass the territory they describe. They use the sheer description of physical detail to delineate not wilderness but a territory, not mere land but a country, and as such they comprise a genre that might usefully be called the "literature of place."

A survey of a group of representative texts will establish the characteristics of the type. Four of these cluster in the 1760s, just at the beginning of the revolutionary period: William Smith's *Historical Account of Bouquet's Expedition against the Ohio Indians* (1765), Major Robert Rogers's *Journals* and *Concise Account of North America* (both 1765), and William Stork's *Description of East Florida* (1769). Two later texts reflect the maturing genre: James Adair's *History of the American Indians* (1775) and Jonathan Carver's *Travels through the Interior Parts of North America* (1781). Three instances of the genre go beyond a fragmentary description of the land to describe what Jefferson calls a "country." These texts were written after independence had been declared, but before its permanence had been secured: William Bartram's *Travels through North & South Carolina, Georgia, East & West Florida, the Cherokee Country, the Extensive Territories of the Muscogulges, or Creek Confederacy, and the Country of the Chactaws; containing An Account of the Soil and Natural Productions of those Regions, together with Observations on the Manners of the Indians* (composed late 1770s–early 1780s; published 1790), Thomas Jefferson's *Notes on the State of Virginia* (composed 1781–82; published 1785), and J. Hector St. John de Crèvecoeur's *Letters from an American Farmer* (composed 1768–81?[2] published 1782).

Two intellectual traditions figure importantly in these texts: the literature of travel and the science of natural history. The travel genre, venerable and enduring, provided the narrative framework on which most of these texts were built. More important, natural history provided these men with a way of looking at the world, with a way of describing what they saw, and with an overarching scheme in which to fit what they had seen. It provided them with method, rhetoric, and context in their descriptions of the new land. Smith, Rogers, Stork, Adair, and Carver used the methods and rhetoric of natural history as one of several different schemes for ordering the information they discovered on their journeys. Bartram, Jefferson, and Crèvecoeur, better trained and established more centrally in the intellectual mainstream of natural history, observed parts of North America with the method of natural history primary, its rhetoric clearly governing significant portions of their texts.

NATURAL HISTORY

Natural history in the eighteenth century was a broad area of scientific inquiry circumscribing the present-day disciplines of meteorology, geology, botany, zoology, and ethnology. It encompassed "the aggregate of facts relating to the natural objects, etc. of a place, or the characteristics of a class of persons or things" (*OED*). One commentator, writing soon after the turn of the nineteenth century, divided the science of the previous century into three distinct eras: through 1735, Newton's physics dominated; 1735–70, natural history lead by Buffon and Linnaeus gained ascendancy; in the last decades of the century, Lavoisier, Priestley, and Cavendish established chemistry as the preeminent science.[3] Since large portions of North America contained little except natural objects (American Indians, as a primitive indigenous population, were counted in that class), the science that focused on this subject matter was a logical guide to exploration. It provided a set of concerns that corresponded exactly with the objects in a new, "uncivilized" country. In addition, it provided a method for investigation and a rhetoric for the verbal descriptions that resulted from that method.

Natural historians employed two basic procedures: collecting and observing.[4] Everything portable was collected—rocks, fossils, mammoth bones, fruits, seeds, roots, leaves, flowers, trees, shells, and animals both dead (insects, stuffed birds) and alive (bullfrogs, snapping turtles).[5] American Indians, along with their artifacts, were sent to Europe. What could not be carried off (like the weather) could be looked at and observations recorded. Collecting and observation were ends in themselves, and an active international community supported these activities.

Bartram, Jefferson, and Crèvecoeur were members of the international community of natural historians. William Bartram was connected to this circle through his work as a plant hunter and illustrator. His patrons included John Fothergill, who sponsored his travels in the American Southeast. Jefferson was himself a plant collector and a patron of plant merchants. He knew the value of exploration to discover new plants and animals and was a methodical experimental gardener and farmer. Crèvecoeur knew plants through his experiences as a farmer in America, as well as through his work as an apothecary or plant hunter.[6] On his return to France, he wrote treatises urging French farmers to cultivate various New World plants, including the potato. Although they were interested in the full range of natural history subjects, for these men botany was primary. In this emphasis they mirrored the scientific community as a whole: the botanical contingent of the natural history circle was by far the most numerous.

Centered in Europe with important outposts in the New World, the botanical community consisted of Old World patrons, of their clients (the New World plant hunters), and of trained botanists working at universities, botanical gardens, or herbaria.[7] The patrons were collectors, many of them devoting considerable money and time to acquiring an extensive collection of exotic plants. They prized new varieties and imported them from all over the world. Their sources for these plants were plant hunters, who were travelers in new lands, or, in the case of America, residents. The collectors wrote out their requests, usually including a blanket statement that anything of interest should be included. The plant hunters foraged for the plants, dug up small specimens, took seeds or cuttings from plants too large to ship whole, packed them, and put them on a boat for Europe. The collector paid for the shipment and planted its contents in his garden. If the plant grew, the patron prepared an herbarium specimen—a cutting carefully chosen for its representativeness. It included seeds, fruit, a flower, and enough leaves to indicate their shape and the pattern of their arrangement on the stem. This specimen was dried, mounted, and sent to an herbarium where a trained botanist compared it with the other specimens in the collection. Following the Linnaean system, he assigned it a name and wrote a description of it. The speciman itself was catalogued and filed. If time and investigation proved that it was truly distinct from each of the other known plants, it could then take its place in the growing list of all the plants in the world. Often the patron, rather than the plant hunter, received credit for the discovery, including the honor of having his name incorporated in the Linnaean nomenclature.

John Bartram (1699–1777), William Bartram's father and teacher, is a figure whose interests, reputation, and career provide a case study in the relationship of European members of the natural history circle to American naturalists. As one of the founders of the American Philosophical Society and an internationally recognized naturalist, he was known to Jefferson, who later served as president of that organization. His friend Crèvecoeur chose him as the subject of Letter XI in his *Letters from an American Farmer.*

John Bartram was a Quaker who farmed 200 acres on the banks of the Schuylkill in Kingsessing, near Philadelphia. There he built a house and established one of America's first botanical gardens. He was a plant hunter, making excursions as far south as Florida, as far north as Canada, and west to what is now Oswego, New York, to collect specimens—primarily plants but also animals and minerals. His patron was British woolen draper Peter Collinson (1694–1768), also a Quaker, and perhaps best known to students today as the correspondent to whom Benjamin Franklin addressed some of his most important scientific and social scientific correspondence.[8] Collinson was successful enough at business to indulge his passion for collecting exotic plants for his garden in Peckham, Surrey. For thirty-four years, ending only at Collinson's death in 1768, he and John Bartram exchanged seeds, cuttings, and information. Their correspondence is instructive of the relationship between a most accomplished plant hunter, which John Bartram certainly was, and the best-meaning of patrons, a frequent judgment bestowed on Collinson.[9]

Bartram's garden was a commercial enterprise; he was in the seed business. In a more subtle way, Collinson's garden was also commercial, not because he sold the fruits of the plants he cultivated, but because it provided him with an entrance into a portion of society that ordinarily had little to do with middle-class Quaker cloth merchants.[10] His concern with exotic plants was not wholly a disinterested scientific one. It is the clash of Collinson's practical acquisitiveness with John Bartram's desire for a formal, scientific way of proceeding that is most instructive about their correspondence. When arduous exploration and collection do not result in the kind of notice that a collector and describer hopes to receive from the scientific community, one alternative open to him is to write a book and have it published separately from the work of his patron, thus providing a competing written record to vie with the official record that the scientific community produces and sanctions. Such an urge to provide a separate record motivated William Bartram when he wrote his *Travels*. The literature of place owes its existence in part to the authors' desire to receive credit for their discoveries.

John Bartram begin supplying Collinson with America's natural productions in 1734. His interest in systematic botany began in 1736 when James Logan, a pioneer in botanical experimentation in America, lent Bartram a copy of Linnaeus's *Systema Naturae* (1735).[11] This volume contained a revolutionary method for identifying plants. Reduced to its essentials, the Linnaean system provided a means for classifying plants first by counting stamens and pistils, then by observing the shape and distribution of leaf, flower, and fruit. The outlines of the system could be (and were) reduced to a one-page chart.[12] Bartram taught himself enough Latin to master Linnaean nomenclature and quickly became adept at applying the new "sexual system" for identifying plants. Soon after, he began to name the plants he supplied to his patron. Collinson found this practice unnecessary. "As to thy care of the names," he wrote, "it does not much signify; for when I see them grow, or flower, can soon distinguish them."[13] In the enterprise to which Linnaeus gave such impetus—the discovery and classification of every natural production on earth—the act of naming is central. The permanent record of botanical discovery preserves the first complete, accurate Linnaean description of a new plant. When, in a misguided attempt to look out for Bartram's best interests, Collinson offered to name a plant the American had sent him, he was also offering to take credit for its introduction into the international botanical community. He believed that Bartram should not delve too deeply into the scientific side of collecting: "A man of prudence will place this to a right account, to encourage thee to proceed gently in these curious things, which belong to a man of leisure, and not to a man of business. The main chance must be minded."[14] The main chance, of course, was Bartram's farm, and the business of collecting and selling, rather than studying and naming, botanical specimens. Collinson's patronizing attitude is most evident when, in 1737, Bartram asked him for a botanical text in payment for a shipment that he had made to him. Since cash payment was often difficult, Collinson usually paid his American employee by sending him goods that could be resold. This time, however, Bartram asked for a copy of Tournefort's *Institutiones rei herbariae*. Collinson refused:

> I shall . . . take notice of thy request to buy Tournefort. I have inquired, and there are so many books, or parts, done, as to come to fifty shillings. The first part may be got, perhaps, second-hand; but the others, are not yet to be expected. Now I shall be so friendly to tell thee, I think this is too much to lay out. Besides, now thee has got Parkinson and Miller, I would not have thee puzzle thyself with others; for they contain the ancient and modern knowledge of Botany. Remember Solomon's advice; in reading of books, there is no end.[15]

Bartram's request is that of a would-be scholar. Tournefort's *Institutiones,* in contrast to Parkinson's *Herball* and Miller's *Gardener's Dictionary,* is a book of theory; it offered not only descriptions of plants, but rules for producing such descriptions. For Collinson, theory lay outside Bartram's proper interests, although the patron introduced his client to many luminaries in the natural history circle, including Linnaeus, Gronovius, Buffon, Sir Hans Sloane, Solander, and Catesby.[16] Yet in his dealings with Bartram he displayed a condescension typical of Old World scientists' attitudes toward their New World counterparts. In botany as in all the sciences, the work of collection—whether of specimens or of "observations that could not be duplicated at some other time and place"—was what Europe most wanted from America.[17] John Bartram established his international reputation as a natural historian both because of and in spite of his relationship with his patron.

The next generation of American natural historians had similar difficulties with regard to patrons. In 1788, before the publication of *Travels,* William Bartram wrote to Robert Barclay, a wealthy brewery owner and amateur botanist, to ask for his help in securing belated recognition from the botanical community:

> I collected these specimens amongst many hundred others about 20 years ago when on Botanical researches in Carolina Georgia and Florida . . . very few of which I find have entered the *Systema Vegetabilium,* not even in the last Edition. . . . These remains with some more that I have kept by me to this time, which I cheerfully offer for the inspection & amusement of the curious, expecting or deserving no other gratuity than the bare mention of my being the discoverer, a reward due for traveling several thousand miles mostly amongs't Indian Nations which is not only difficult but Dangerous, besides suffering sickness cold & hunger.[18]

When the Revolution intervened, independent written observation replaced epistolary communication with England and the Continent. Native plants, once catalogued and named by foreigners, were chronicled by native writers. They fashioned an accounting, a listing, a declaration of America's natural productions for the same audience—the candid world—to which the Declaration was directed. The declarations were complementary, two kinds of definition of the new land. The political Declaration began the process by which the country would be written into being. The natural historical declarations defined the place where that country would exist, and named and illustrated the objects of the creation that would furnish the new land. America's nature was no longer in the hands of Europeans. The enforced interruption of publishing caused by the Revolution actually advanced the war's aims—America's independence extended to her natural history authors, who were left on their own to write representations of their new country.

The relationship between the American and European contingents of the natural history circle might have suffered a breach, but individual members retained their allegiance to the principle of order that had made it possible for natural historians remote from centers of learning to do natural history: the Linnaean system. Recognition in the natural history circle required the mastery and application of Linnaeus's revolutionary system of nomenclature. This system provides one key to how its practitioners, including William Bartram, Jefferson, and Crèvecoeur, looked at the world, how they arrived at the conclusions they did, and why they wrote in the style that they did.

Linnaeus revolutionized natural history, and especially botany, by reforming its language. Until his time there was no universally recognized method for classifying and naming plants. Descriptors multiplied as more of a given plant's characteristics were added to its scientific name in an effort to distinguish it from neighboring species. The names of plants could be many words long and could vary from authority to authority. Linnaeus changed all that. In so doing, he made the conception of the Great Chain of Being completely rational. After Linnaeus, the binomial label pointed both to a species or kind unique from all others, and to a link on the Chain. It evoked, at once, the very specific and the whole order of the universe.

The landmark work that established the ascendancy of Linnaeus's language for natural history was the *Systema Naturae*, first published in 1735. Linnaean natural history was a science of description. Modern biology, which concerns itself with the inner workings of plants and animals, was to be built upon the descriptions that poured forth after Linnaeus regularized natural historical nomenclature. However, at the time of Linnaeus, concern with the inner workings of organisms was eclipsed by concern with naming and thereby classifying them based on their surface appearances.

Linnaeus's "system of classification, based, in the botanical field, upon the differences of the sexual organs of plants" was an ideal one for an American to learn and use. It was easy to remember. It was reduced to a one-page chart and often amounted to little more than counting stamens and pistils.

Armed with the chart and a magnifying glass, an American could participate in the most important scientific endeavor of the day. Its language was simple. Linnaeus's introduction of binomial nomenclature "put a final end to the hopeless length and complexity of botanical and zoological specific names, and sharply differentiated the naming of organisms from the description of them." Most American naturalists worked in the field rather than in botanical gardens. The Linnaean reduction of natural history's nomenclature made their tasks easier. It was universal. Linnaeus drew up "a set of terms, each with clearly defined meaning, for designating concisely the distinguishable parts and organs of plants, and the several types of form of which each part is susceptible."[19] The universal acceptance of these terms meant that an American plant hunter could build a verbal description of a native plant that would mean the same thing to any member of the natural history circle anywhere on earth. Despite their remoteness from the Old World centers of natural historical learning, Americans, equipped with their eyes, with Linnaean nomenclature, and with a pen, could make a contribution. When remoteness became inaccessibility—when the Revolution cut American plant hunters off from their largely British patrons—they could carry on without contact with European seats of learning, knowing that the descriptions they produced and the names they assigned were universal, sanctioned not by the authority of experts, but by the method that produced them.

If Linnaeus offered a different universal name for every natural production, he also offered a reaffirmation and an updating of the overall scheme that embodied all of these productions—the Great Chain of Being.

> Every discovery of a new form could be regarded, not as the disclosure of an additional unrelated fact in nature, but as a step towards the completion of a systematic structure of which the general plan was known in advance, an additional bit of empirical evidence of the truth of the generally accepted and cherished scheme of things. Thus the theory of the Chain of Being, purely

> speculative and traditional though it was, had upon natural history in this period an effect somewhat similar to that which the table of the elements and their atomic weights . . . had upon chemical research.[20]

Thus, America had, in Linnaean nomenclature, a language that could be accurately applied to it and an overriding scheme into which its productions could fit. Amid the chaos of the Revolution, American natural historians were contributing to this cosmic order when they described and named the productions of their country. They used the language of natural history to make America intelligible to readers who had never seen any part of the Western Hemisphere. The New World of exotic mystery, of distance-shrouded indistinctness, gives way to a sharp-edged, delineated, concrete description systematically and rationally related to the Old World.

Michel Foucault has described the method that field natural historians, specifically botanists, would have used in applying Linnaeus's system.[21] Choosing one plant in a given locale, an observer described it completely, omitting nothing. For the next plant, the observer listed only those characteristics that differed from the first plant. For the third, the observer listed only characteristics that differed from those of the first two, and so forth. "So that, at the very end, all the different features of all the plants have been listed once, but never more than once. And by arranging the later and progressively more sparse descriptions around the earlier ones, we shall be able to perceive, through the original chaos, the emergence of the general table of relations."[22] A version of this method, abbreviated by past experience and by the established work of one's botanical forebears, would be set into motion by an encounter with a strange plant. A description that classified the new plant within the "general table of relations" already established would result.

William Bartram had such an encounter with the plant that he would name *Franklinia*. During his first trip south, William and his father found on the banks of the river Altamaha in

Georgia what John described in his diary as "several curious shrubs."[23] But it was autumn, and the plant was not in bloom nor was it bearing fruit. When William returned alone in a subsequent summer he saw the plant "in perfect bloom, as well as bearing ripe fruit."[24] In 1788, some years after his journey, but before the publication of the *Travels*, he sent his drawing of the plant, made from a living specimen (presumably growing in his family's garden) to Robert Barclay with the letter of appeal for botanical recognition quoted earlier. Barclay labeled it *Gordonia pubescens*.[25] Bartram, however, was convinced of the tree's singularity, and he argues with Barclay's identification in the chapter of the *Travels* devoted entirely to this plant:

> On first observing the fructification and habit of this tree, I was inclined to believe it a species of *Gordonia*; but afterwards, upon stricter examination, and comparing its flowers and fruit with those of the *Gordonia lafianthus*, I presently found striking characteristics abundantly sufficient to separate it from the genus, and to establish it the head of a new tribe, which we have honoured with the name of the illustrious Dr. Benjamin Franklin. *Franklinia Alatamaha*.[26]

"Comparing its flowers and fruit with those of the *Gordonia*": this is the method at work, the process of comparing one known, named plant with an unidentified, unnamed one to discover relationships. The goal is to insert it into the table of those relationships. That table was extensive. The first edition of Linnaeus's *Species Plantarum* (1753) ran 1200 pages; it is nothing more (or less) than a list of every known plant with sufficient descriptors to permit an investigator to find for any new plant its proper place in the list.

The Linnaean method led to a literary style and to a rhetoric. It led to a tacit consensus among natural history practitioners, including Bartram, Jefferson, and Crèvecoeur, about how sentences ought to be written and how discourses ought to be constructed. An obvious form for recording natural historical observation is the list. One of the earliest American instances of this style and rhetoric was written on the frontier and read before the Royal Society on 21 May 1772: John Reinhold Forster's "Account of Several Quadrupeds from Hudson's Bay." The secretary to the society appended this note to the paper's title:

> From the factory at Hudson's Bay, the Royal Society were favoured with large collections of uncommon quadrupeds, birds, fishes &c. together with some account of their names, place of abode, manner of life, [and] uses. . . . The descriptions contained in the following papers were prepared and given by Mr. Forster, before his departure on an expedition, which will probably open an ample field to the most important discoveries.[27]

The paper consists of a series of numbered entries, the first of which begins in this way:

> 1. ARCTIC FOX, Penn. Synops. of Quadr. p. 155. n. 113.
> *Canis Lagopus,* Linn.
>
> Severn River.
> A most beautiful specimen in its snowy winter furr; this animal seems to be lower on its legs than the common fox, and is prodigiously well secured against the intense cold of the climate, by the thickness and length of its hairs, which are at the same time as soft as silk.

It concludes, "The specimen sent is full grown, and its furr quite in season."[28] The form of Forster's account is a commentary accompanying an actual physical specimen. He does not need to describe the animal for the reader. Presumably, at the reading of the paper, the specimen would have been visible to the assembled listeners. And this species was not new. The reference to Pennant's *Pennsylvania Synopsis of Quadrupeds* and the Linnaean nomenclature, *"Canis Lagopus,"* attest to that. Forster was writing to confirm a description, a name, and a set of characteristics.

When he does not agree with a prior description, he writes a different sort of note:

> 6. SKUNK, Penn. Syn. Quadr. p. 233. n. 167.
> Kalm's Travels, I. 273. tab I.
>
> It answers to Mr. Pennant's description, except that the white stripe on the head is not connected with that on the back, and that the brown area, which is left between the two white stripes on the back, is broader than he describes it.[29]

Here, after the usual reference to the authorities, in this case Pennant and Kalm, the only thing to add is the new observation about the stripes.

At its most primitive, natural history writing is description to accompany actual specimens. Words and things are linked by the actual presence of the thing described. Its form is the annotated list. The Linnaean names, the references to earlier authorities, and the appended observations of the animal all serve to identify and describe an object that is present when the entry was being read. A more extreme sort of list, one that is unannotated, is the catalogue, such as Forster's *Catalogue of the Animals of North America. Containing An Enumeration of the known Quadrupeds, Birds, Reptiles, Fish, Insects, Crustaceous and Testaceous Animals; many of which are New, and never described before.* Here is what its first few entries look like:[30]

CLASS I. QUADRUPEDS.

DIV. I. Hoofed.

SECT. I. Whole Hoofed.

Genus.			syn. quad.
I. HORSE	Generous	*E.*	No. 1
	Ass	*E.*	3

"E.," Forster tells us, means that the animal so noted is of European origin. The "syn. quad." column refers to the corresponding

item in Pennant's *Synopsis of Quadrupeds*. Lists of this kind, and this one runs twenty-eight pages, were a means to argue the placement of a difficult discovery, a way to gather authorities together in an up-to-date version of the information. They were also the written representation of the Great Chain itself. How could one argue that no such chain existed when the chart or the table seemed a visual representation of it, showing clearly that there was such a thing to demonstrate, to portray? To an eighteenth-century reader, it could not be a representation of nothing.

During his trip to the southeastern territories, William Bartram sent his patron, Dr. John Fothergill, 209 herbarium specimens. They were usually accompanied by annotations, but sometimes, as in the case of *Myrica inodora* Bartr., they simply consisted of the dried plant itself, stitched onto a leaf of paper. This particular specimen is the type specimen, that is, the one to which all other plants are compared to make an identification. It was described in *Travels*.[31]

> In my excursions about this place, I observed many curious vegetable productions, particularly a species of *Myrica* (*Myrica inodora*): this very beautiful evergreen shrub, which the French inhabitants call the Wax tree, grows in wet sandy ground about the edges of swamps, it rises erect nine or ten feet, dividing itself into a multitude of nearly erect branches, which are garnished with many shining deep green entire leaves of a lanceolate figure; the branches produce abundance of large round berries, nearly the size of bird cherries, which are covered with a scale or coat of white wax; no part of this plant possesses any degree of fragrance. It is in high estimation with the inhabitants for the production of wax for candles, for which purpose it answers equally well with bees-wax, or preferable, as it is harder and more lasting in burning.[32]

This is not the usual herbarium specimen description. It is located in space and time—in narrative, although the place where an herbarium specimen is found will often be identified. This is a more typical description: "13. An inhabitant of our

Savanahs grows 4 or 5 feet high bearing numbers of small blue flowers in long slender spikes."[33]

Forster and Bartram both engage in the most common, most minimalist means for representing nature—that is, they do not describe it; rather, they send the specimen along and annotate it, or simply label it. The list, the chart, the label are the natural historical equivalent of the formula in physics—the tersest, most reductive expression possible. In physics, a phenomenon has been reduced to its most essential components, and these components have been employed in an expression, the formula, that defines their relationships to one another. In the case of the naturalist's label, one of two things is happening.

For a specimen whose type characteristic is unknown or almost certainly in question, the label describes a relationship between a word and a thing that is as yet impermanent. The herbarium researchers will be able to establish the propriety of the relationship between the Linnaean label and the specimen. Either the name will be associated with the thing or it will give way to a binomial with a better claim.

For a type specimen (the flower, leaves, seeds, and fruit of a plant that is a separate species from all others), the identity of a given kind of thing is being fixed forever, certainly within the community of adherents to the system, and, a Linnaean would argue, within the natural world as well. An individual is being nominated to serve as the type or the representative of all other members of that species of thing. A name is assigned that applies to a member of that species and to none other. The thing referred to, as well as the kind of thing referred to, is fixed for all time. This process of determining the relationship between a name and the thing named forces language to be more precise than it is usually thought capable of being. The problems encountered when the referent and the thing referred to are separated by space, and thus susceptible of not being associated in the way the referrer had intended, are eliminated.

These Linnaean names were the passwords to botanical universality, guaranteeing that readers everywhere would take these names as referring to the same plants and animals. On the level of diction, then, the style of these books is partly determined by their authors' scientific backgrounds. Linnaean diction provides a way to speak to anyone anywhere about what is here. The Linnaean list shows the similarities of one American thing to any number of things from any number of places. A thing's placement on the list relates it to any number of natural history productions in any of the explored areas of the world. Thus for European readers, America seems at once strange and familiar.

A Linnaean specimen entry begins with the generic name, then offers the "specific differential character," the elements of the plant that distinguish it from others in the genus. A third part is the trivial name, which, with the generic name, makes up the two Latin names that are entered in the lists. A fourth part is the plant's habitat, and last is a brief description or annotation, or synonyms, often with references to earlier systems or illustrations of the plant in question.[34] Here is an example of an entry from Linnaeus's *Species Plantarum:*

noli tangere. 7. IMPATIENS pedunculis multifloris solitariis,
foliis ovatis, geniculis caulinis tumentibus.
Fl. suec. 722. *Dalib. paris* 270.
Impatiens pedunculis solitariis multifloris.
Hort. cliff. 428. *Roy lugdb.* 431.
Hall. helv. 405. caule angulato.
Gort gelr. 502.
Balsamina lutea f. Noli me tangere. *Bauh. pin.* 306.
Noli me tangere. *Col. ecphr.* I. *p.* 49. *t.* 150.
Habitat in Europae, Canadae *nemoribus.*

This entry appears on a page under the running head "Syngensia Monogamia."[35] The Linnaean universe was divided into classes, orders, genera, and species. "Syngesia" is the Linnaean name for the nineteenth class. Impatiens have nineteen male organs, or stamens. "Monogamia" names the order. Impatiens have one female organ, or style.

The first paragraph of the entry, its parts labeled and translated, reads:

noli tangere	Name of species.
IMPATIENS	Name of genus. The binomial name of this plant is *Impatiens noli tangere*.[36]
pedunculis multifloris solitariis	Descriptor. Many flowers, single on footstalks.
geniculis caulinis tumentigus	Descriptor. Nodes of the stem swollen.
Fl. suec. 722	Listed in Linnaeus's *Flora Suecica*.
Dalib. paris 270	Listed in Thomas-Francois Dalibard's *Florae Parisiensis prodromus*.

Subsequent paragraphs offer references to other authorities, with the various names that those authorities had given this plant. The final paragraph identifies the plant's habitat, both by location, Europe and Canada, and by the kind of locale in which an observer might see the plant, "nemoribus," in woods and groves.

The Latin of the system is a shorthand, derived from medieval Latin regularized. One notable feature of the short hand, for our consideration of these authors, is its absence of verbs.[37] Linnaean Latin, the lingua franca among naturalists, is deliberately static. The unexpressed, unwritten verb implied in all of the descriptions is "to be." In the lists, things merely are. All possible predicates are gone: reproduction, nutrition, sensitivity, or movement, all unstated (and thus unimportant) before the regularizing litany of the names and the descriptors.[38]

The Linnaean method enabled discovery through observation but, like all methods, defined what would be observed and how. The characteristic way of thinking fostered by the Linnaean method emphasizes similarities among individuals belonging to a single group and differences between a given group and all other groups. In applying this method to plants, an observer learns to call each group by a separate, universal name. Individual plants share essential characteristics with other members of their group or species. Confusing plenitude is resolved into the beauty and order of a list.

In the *Systema Naturae* Linnaeus included man in his list of animals. He was the first to take this step. After Linnaeus identified the ape as a close relative of man "the idea of a hierarchy and a continuity in nature, of a scale and a chain, gained in popularity." If the links of the chain were thus strengthened, they were also added to, as Linnaeus divided man into more than one species: below *Homo sapiens,* he placed *Homo troglodyte, H. caudatus,* and *H. lar.*[39] He was also the first to incorporate racial distinctions into his system. In the 1758 edition of *Systema Naturae* Linnaeus offers "the first formal definition of human races in modern taxonomic terms," *Homo sapiens europaeus* and *Homo sapiens afer.* Whites and blacks were of the same species, but of different races. The newness of this classification of man with the animals, the linking of *Homo sapiens* with fanciful creatures for whom there was scant evidence (the supposedly cave-dwelling *Homo troglodyte*) and the division of man into separate races began the speculation that would culminate, as early as 1766, in Hume's claim that non-white races were of a different species from the white race.[40] The inclusion of man in the Linnaean lists made him the same target for observation that other natural objects were. "[M]an himself now had history, and this history no longer consisted solely of the customary genealogies starting with Adam."[41] The division of man into separate races—a natural enough development in a system that emphasized differences—made it likely that observations would proceed along racial lines. Thus, although nominally "man" had become a subject of investigation, in practice men of other races, perhaps of other species, became the objects of this scrutiny. This addition of man to the natural historical list of subjects occurred during a period of intense botanical exploration and discovery. Thus the methods of exploration and description applied to man in America at this time were the same as those applied to plants. Explorers looking for exotic, native American productions found that American Indians fit these criteria admirably.

In applying this method to human beings, an observer likewise learns to recognize each group by a set of characteristics that are by definition shared with other individuals of their group. The qualities of a particular individual—personality, even sex—are overlooked by the explorer because the method overlooks them. The representations of human beings produced by the method are now called "manners-and-customs" descriptions, after the topics on which they usually focus. All of the authors under consideration produced such descriptions of America's native peoples. Mary Louise Pratt notes that these manners-and-customs descriptions, the antecedent of ethnography proper, "codify difference" and "fix the Other [the people under observation, in this case, American Indians] in a timeless present." Travel writing, Pratt notes, places these manners-and-customs descriptions in a context where they are "in play with other sorts of representation," especially narrative. She describes a feature of nineteenth-century travel writing that applies to the strict practitioners of the method described above: "Indigenous peoples are relocated in separate manners-and-customs chapters as if in textual homelands or reservations, where they are pulled out of time to be preserved, contained, studied, admired, detested, pitied, mourned."[42] This exile within the text is common in natural history descriptions. A native American who existed as an individual with a unique personal history, with his or her own memories, was grouped with the American Indians. All characteristics were either shared or nonexistent, at least as far as the system could discern and record those characteristics.

This leveling of individuality in natural historical descriptions of human beings met with various responses. Bartram struggled with it, attempting to represent American Indians in narrative but ultimately placing them firmly in their own section of *Travels*. Jefferson ventured one American Indian narrative, then eliminated its narrativeness through legalistic and obsessive repetition. Only Crèvecoeur managed to include native Americans in his narrative, in the form of the imaginative projection of one of his narrators, James, who foresaw his own "transmutation" into an American Indian when he abandoned his farm at the onset of the Revolution.

Time is the essence of narrative, but Linnaeus's scheme is timeless. For him "there is not and cannot be even the suspicion of an evolutionism or a transformism . . . for time is never conceived as a principle of development for living beings in their internal organization."[43] To find their location in the table was to take the plant, or the person, out of context. It put them into the grand scheme, located their links on the Great Chain. It also took them out of time, out of the narrative context in which their describers found them, trading the particularity of time and place for the atemporal, ageographic generality of the scheme.[44] Time and space suspend before the eternalizing, location-dissolving calling of the Linnaean names.

Practitioners of this method journey into a world whose order can be reduced to items in a table, whose plants and peoples are related not by the local constraints of neighborhood but by the global constraints of similarity, of relatedness through visible elements of roots, stem, leaves, and especially flowers and fruit. The system ordains that they will observe what others have found—plants with leaves of such-and-such a shape, with flowers of a certain type, fruit of a given kind. Their eyes have been educated by the Linnaean system and the inductive method that underlies it. What is less ordained, less noticed in the system, is where they will find what they find. Journeying to unexplored areas is a way to break free of the table, to get beyond the known. It is also a way to introduce the temporal into what Foucault has described as "the gradation of beings" into which "the temporal series cannot be integrated."[45] The basic elements of travel—the separation of day and night, the narrative reality of eating, riding, observing, sleeping, leaving, and returning—provide what the system ignores: time. And in writing texts that depend to varying degrees on narrative, authors of the literature of place bracket with their own actions the frequent lists, the pauses for description of the soil and of the inhabitants.

The two modes, narrative and natural historical description, typically occupy discrete passages in these texts. Narrative suspends. Description replaces action. Natural historical representation presents America at its most characteristic—its unique plants, animals, peoples, and scenes—and as outside of time. America seems new; it is a place where events have not intruded. Native Americans are subsumed under this natural historical description, becoming entries on a list, links on the chain. The rhetoric of this description denies them any history, individual or cultural, because that rhetoric did not include a way to represent time.

THE LITERATURE OF PLACE

Linnaeus's reform of natural history provided the intellectual justification and method for the books of Bartram, Jefferson, and Crèvecoeur, as well as providing a rhetoric for long passages in their texts. Had the organization of these books been based on Linnaean models, the authors would have written what Linnaeus himself was best known for—a book-length list. Or they would have written what Buffon did in his *Histoire Naturelle*—natural history essays. Their books do include lists and natural history essays, but in a structure that owes much to the long tradition of travelers' and explorers' accounts of the new land. Since Columbus such accounts have informed Old World readers of the characteristics of the New World, one text venturing on unreported ground, subsequent texts refining, correcting, and extending the first accounts of what a given place consists of—what soil, mountains, and rivers compose the territory, and what is to be found within that territory.

Narrative is the contribution of travel literature to the literature of place. It is an essential feature of the travel genre. Despite its importance, narrative does not overshadow description. The "two most-frequently cited classical models" for the eighteenth-century travel account were Pausania's *Graeciae descritio* and Horace's "Journey to Brundisium," whose descendants are, respectively, natural histories and memoirs. Eighteenth-century travel literature accomplished a "blending" of the dominant mode of discourse in each form—description in natural histories and narrative in memoirs. Samuel Johnson reflected this balancing of the two modes when he described travel literature as "Science . . . connected with Events." Joseph Addison's *Remarks on Several Parts of Italy* (1705), one of the most famous and popular instances of English travel literature, served as a guidebook, both for travelers and travel writers.[46]

Addison's *Remarks* provides a ready contrast to the typical colonial American travel narrative. Addison travels in order to disapprove; the travelers in the American tradition travel to discover. Addison covers ground that had been visited and written about for centuries; the Americans travel to less-visited areas. Addison travels through civilized territory; the Americans travel through barely inhabited regions. Addison travels chiefly to see monuments of human history and to be reminded of the literary past; the Americans travel to observe the natural history of the regions they visit.

Eighteenth-century American travel narratives often retain the characteristics of their occasional beginnings. Captain Harry Gordon, chief engineer of the Western Department in North America, kept a journal of the trip he took in 1766 down the Ohio, the Illinois, and the Mississippi rivers to New Orleans and into West Florida. He acted as surveyor, measuring and recording the breadth of the rivers, while scouting the riverbanks for likely trading post sites.[47] In 1780, Bishop Frederick Reichel of the Moravian church recorded the details of his trip from Lititz, Pennsylvania, to Salem, North Carolina, noting the cost of ferries and the names of the hymns the travelers sang.[48] For some New World travelers, like the engineer, journal writing was a professional duty. For others, like the bishop, it was the habit of a literate man. Intended only to record measurements made or events experienced on a given day of the trip, these accounts are unembellished by introductions, statements of intention, or other readerly amenities.

Other colonial travelers wrote accounts whose scope and execution raise them above the common, workmanlike level of the majority: Major Robert Rogers's *Journals* and *Concise Account of North America* (both 1765), William Smith's *Historical Account of Bouquet's Expedition against the Ohio Indians* (1765), William Stork's *Account of East Florida* (1766), James Adair's *History of the American Indians* (1775), and Jonathan Carver's *Travels through the Interior Parts of North America* (1778).[49] In scope, they all attempt generalizations about a given territory or people. They seek the characteristic, the representative in their depictions of terrain, plants, animals, and men. In execution, each is written by a man who intended to publish. Their primary audience was the British public, but Americans by the time of the Revolution were also reading these texts to learn about remote areas of their own continent.[50] Rogers, Smith, Stork, Adair, and Carver each announce the same intention for writing his text—to be of use. Each exploits a wide range of forms and kinds of discourse to illuminate two broad subject areas—the nature of the land itself, and the nature of the original inhabitants of that land. Bartram, Jefferson, and Crèvecoeur write of the same desire to be of use and employ the same forms and discourse in pursuit of the same subjects. The earlier texts attest to the continuity of intention and illustrate the range of formal and stylistic options each writer could call upon to convey his information, while they foreshadow certain elements of the three later works. They also point to the consequences of the choice their authors made between narrative and description, consequences that are particularly acute in the representation of America's natives.

The earlier writers have a common goal in writing. In his *Journals,* Rogers sets aside the "private views" he might have for publishing his soldier's diary of campaigns in the French and Indian War, to "claim the merit of impartially relating matters of fact, without disguise or equivocation." These matters of fact, he goes on to claim, might prove useful: "[S]hould the troubles in America be renewed, and the savages repeat those scenes of barbarity they so often have acted on the British subjects, which there is great reason to believe will happen, I flatter myself, that such as are immediately concerned may reap some advantage from these pages."[51] His *Concise Account* is arranged topically, by locale; in it Rogers sounds the theme that underlies the production of most works in the genre: "The only thing I mean to do is, in a simple and intelligible manner, to relate such matters of fact as may be useful to my country. . . ."[52] William Smith, listed on the title page of his *Account of Bouquet's Expedition* as "a Lover of his Country" bases his history of Colonel Bouquet's campaign against the native inhabitants of the Ohio River valley on Bouquet's own notes and dispatches. Bouquet was made a brigadier general on the strength of his successful campaign but died soon after in Pensacola, Florida. In writing his account of Bouquet, Smith seeks to secure his subject's place in history, while at the same time offering a model for dealing with the American Indians: "The Colonel's firm and determined conduct in. . . the campaign . . . shews by what methods these faithless savages are to be best reduced to reason."[53] Within the larger goal of writing Bouquet's history, Smith finds instructional value in his treatment of native Americans. Stork, in his dedication to Charles, marquis of Rockingham, assures him, "I have no vices in publishing the following sheets, but the benefits and advantages Great Britain may reap." He later asserts that his "design" in writing is to "make the nation [Great Britain] acquainted with that country [East Florida]."[54]

Adair intended his *History* to replace the "romance and a mass of fiction" previously published about the American Indian. Yet his goal was ultimately to be of use: "My grand objects, were to give the Literati proper and good materials tracing the origin of the American Indians—and to incite the higher powers zealously to promote the best interests of the British colonies, and the mother country."[55] Carver imagines two different kinds of readers, adding amusement to the main theme:

> To those who are interested in the concerns of the interiour parts of North America, from the contiguity of their possessions, or commercial engagements, [*Travels*] will be extremely useful, and fully repay the sum at which they are purchased. To those, who, from a laudable curiosity, wish to be acquainted with the manners and customs of every inhabitant of this globe, the accounts here given of the various nations that inhabit so vast a track of it, a country hitherto almost unexplored, will furnish an ample fund of amusement and gratify their most curious expectations.[56]

The three later writers had similar goals in writing. Bartram hopes to present "new as well as useful information to the botanist and zoologist."[57] Jefferson wrote in response to questions on Virginia submitted by François Marbois, who, as secretary of the French legation, solicited information about the colonies for use by his government. In a letter to Jefferson, Crèvecoeur reveals that he once hoped for a second edition of his *Letters*, and asks for Jefferson's help, explaining "it is not vanity that Inspires it, but a desire that the Second Edition might be more usefull and more correct than the first."[58]

In Rogers's *Journals* and *Concise Account*, Bartram, Jefferson, and Crèvecoeur could have seen illustrated a range of formal and stylistic options available to the writer who has set himself the task of describing a new "country." Yet the contrast between the two texts is striking. The *Journals* is first person; *Concise Account* is third. *Journals* is narrative; *Concise Account*, exposition, chiefly description. *Journals* presents a description of the Great Lakes region to serve as a setting for the events of the narrative. *Concise Account* is all setting, description being its main purpose.

Rogers was a major in the colonial rangers who fought the

French and Indians in the Great Lakes region between 1755 and 1761. His *Journals* is a military diary depicting the movements and engagements of the troops he commanded. Interspersed are letters other soldiers wrote to Rogers, the whole text thus providing a series of first-person narratives. Rogers presents the area around the Great Lakes in occasional glimpses of description set into narrative accounts of his military exploits: "I found the soil near this river [called by the American Indians the Grace of Man] very good and level. The timber is chiefly oak and maple, or the sugar-tree."[59] The trees growing on the land formed a shorthand for determining the land's agricultural worth, the standard of value that measured "use." The best land, according to this system, supported oak, among other species; thus Rogers's mention of the trees, and especially the oak, is an indication of the land's worth, a signal to his contemporaries who would understand the significance of that tree growing on that land.[60] Description of the land, however, is overshadowed by narrative.

Rogers and the correspondents whose letters he includes recount stories of what it is like to live and fight on the frontier. In the midst of describing troop movements and skirmishes with the American Indians, Rogers includes rules for his rangers, which reflect his knowledge of the native Americans' habits: "At the first dawn of day, awake your whole detachment; that being the time when the savages chuse to fall upon their enemies, you should by all means be in readiness to receive them." One correspondent recounts his efforts to return to Fort Edward after Rogers's war party, of which the correspondent was a member, was obliged to leave him behind during an engagement: "Our snow-shoes breaking, and sinking to our middle every fifty paces, the scrambling up mountains, and across fallen timber, our nights without sleep or covering, and but little fire, gathered with great fatigue, our sustenance mostly water, and the bark and berries of trees."[61] In contrast to Rogers's confident movement through the wilderness, this writer and his friend become lost, finally viewing their discovery of a French fort and their capture by the enemy as a deliverance from death at the hands of an insane guide and the wilderness. The reader of this narrative contrasts it with Rogers's own successful travels through these territories and is left with a sense of America's interior as a place where knowledge of the land is vitally important, where landmarks can be recognized by some Europeans, but by all native Americans. It is a place that can turn into an enemy, where any journey can become a circular path on which avoiding death is the paramount concern: "We made a path round a tree, and there exercised all the night, though scarcely able to stand, or prevent each other from sleeping. Our guide . . . straggled from us, where he sat down and died immediately."[62]

In this book Bartram could have seen a narrative depicting movement through the wilderness, which forms the narrative spine of his depiction of the Southeast. Jefferson could have read brief samples of third-person description of America. As a model for his *Letters,* Crèvecoeur could have seen the epistolary form employed to describe the land. Of course, Bartram could have read any of a number of travelers' narratives. Jefferson and Crèvecoeur could have seen models for their works in a number of texts. But the similarities to be found between Rogers's *Journals* and the works of Bartram, Jefferson, and Crèvecoeur demonstrate a continuity of tradition within a formally and stylistically diverse group of texts.

A far less dramatic representation of the interior emerges from *Concise Account of North America,* in which the Rogers of the *Journals* steps back to survey the whole. The *Concise Account* is pure exposition, third person, and impersonal, which turns the landscape into something observed rather than something interacted with, as it was in the *Journals.* The book subdivides America into the coastal colonies, already civilized, and the river valleys of the interior, inhabited only by American Indians. Each section is devoted to a single colony or territory and follows roughly the same organization. In the Virginia chapter, for example, the boundaries of the colony, its history, the soil and climate, the chief exports, and the government are each in their turn described.[63] In a text like this, Jefferson could see a model for the organization of his entire book. The order of the sections in *Notes* roughly parallels the order that Rogers imposes on each chapter of his *Concise Account*. When he comes to describe the unboundaried interior territories, Rogers uses the primary rivers to define an area and provide a name for each section; then he describes the soil and the inhabitants. Implicit in the careful organization of the observations and the objectivity of the third-person exposition is a representation of an orderly territory awaiting civilization. Rogers makes this point explicitly as well: "In a word, this country wants nothing but that culture and improvement, which can only be the effect of time and industry, to render it equal, if not superior, to any in the world."[64] In a text like this, Bartram, Jefferson, and Crèvecoeur could see the third-person description of a given colony or territory, as well as the attempt to offer the reader not only a description of the better-known colonies but also of the little-known territories. In the *Concise Account,* Rogers consolidates existing information and includes new, firsthand observations, a procedure that Bartram, Jefferson, and Crèvecoeur all followed.

Rogers's *Journals* and *Concise Account* illustrate the choice confronting the writer of a text that is intended to represent a given territory, particularly the inhabitants of that territory. In a final section of *Concise Account* separate from the chapters on the colonies and territories, Rogers presents a manners-and-customs description of the American Indians. In *Journals,* where his primary aim is to relate a series of events, Rogers presents the landscape, plants, animals, and native Americans in the same kind of discourse—narrative. In his *Concise Account,* where his primary aim is to provide a verbal survey of America, Rogers places his descriptions of the animals (quadrupeds) and the American Indian in separate sections. A choice of one kind of discourse over another has definite consequences for a given book's representation of America and her inhabitants. In *Journals,* American Indians are not presented as new or even exotic. Rogers understands them in the same way that any person understands a dangerous, but unquestionably human, foe. In *Concise Account,* however, American Indians are presented as exotic and unknown. Elaborate descriptions of them have to be offered in a separate section, whose mode of discourse is not narrative and whose essence is not the relation of events in time; rather, the mode of discourse is atemporal description. Description removes America's natives from context, making them Other. This conflict between narration and description recurs in Bartram, Jefferson, and Crèvecoeur. Each writer struggles with the consequences of choosing a particular mode of discourse, as each writer's account of the territory he sought to represent is constructed from modes of discourse that permit or exclude the representation of action. This is a particular problem in the treatment of American Indians, and each writer makes a decision that places his practice in accord with either Rogers's *Journals* (where the depiction of native Americans is part of the narrative of the rangers' campaigns) or with Rogers's *Concise Account* (where the depiction is a manners-and-customs description that removes American Indians altogether from a narrative and, hence, from an historical context). The American Indians posed the most serious challenge to these authors' powers of representation because there was, finally, no satisfactory choice of discourse with which to convey them and their cultures to a reader.

The remaining four texts each represent a part of America in the manner of *Journals* or *Concise Account.* In addition, each depicts the American Indian from a distinct viewpoint, representing different events, customs, or objects, thus highlighting different themes. Taken together, they provide an anthology of possible themes in the objective depiction of the American Indian. Bartram, Jefferson, and Crèvecoeur might have seen and used such models in their own depictions of the natives they encountered.

William Smith's *Historical Account of Bouquet's Expedition against the Ohio Indians,* like Rogers's *Journals,* is a military diary. It tells the story of the westward movement of troops led by Bouquet, who had been ordered to march into American Indian territory and obtain a peace treaty with the inhabitants. He had proved himself a canny fighter in his victory at Bushy Run and was now to press the advantage obtained in that battle. The march itself was uneventful, primarily because Bouquet had abandoned his European-based open ground order of march for a procedure more appropriate to the American woods. Smith recounts the day-by-day movement of the major and his army beyond Fort Pitt into what is now southeastern Ohio, where Bouquet parleyed with the native Americans, imposing as a condition for peace that all white captives be surrendered by their captors. Smith then breaks into the military and diplomatic narrative to describe the scene as white captives were reunited with their relatives in Bouquet's camp:

> And here I am to enter on a scene, reserved on purpose for this place, that the threads of the foregoing narrative might not be interrupted—a scene, which language indeed can but weakly describe. . . .
>
> The scene I mean, was the arrival of the prisoners in the camp; where were to be seen fathers and mothers recognizing and clasping their once-lost babes; husbands hanging around the necks of their newly-recovered wives; sisters and brothers unexpectedly meeting together after long separation, scarce able to speak the same language, or, for some time, to be sure that they were children of the same parents![65]

The military narrative had no place in it for these domestic reunions, which were the fruits of the success of Bouquet's march into American Indian country.

In addition to describing a scene inappropriate to his military and diplomatic history of Bouquet's campaign, Smith reveals a different side of the American Indians from the "faithless savages" whose promises previously had proven so unreliable: "The Indians too, as if wholly forgetting their usual savageness, bore a capital part in heightening this most affecting scene. They delivered up their beloved captives with the utmost reluctance; shed torrents of tears over them, recommending them to the care and protection of the commanding officer." Smith does not fail to draw the moral:

> These qualities in savages challenge our just esteem. Cruel and unmerciful as they are, by habit and long example, in war, yet whenever they come to give way to the native dictates of humanity, they exercise virtues which Christians need not blush to imitate. When they once determine to give life, they give every thing with it, which, in their apprehension, belongs to it. . . . No child is otherwise treated by the persons adopting it than the children of their own body.

He also does not fail to note the reluctance of some children to part with their captors: "Having been accustomed to look upon the Indians as the only connexions they had, having been tenderly treated by them, and speaking their language, it is no wonder that they considered their new state in the light of a capitivity, and parted from the savages with tears."[66] Smith's representation of the American Indians is split between the "faithless savages" and the possessors of "virtues which Christians need not blush to imitate." The depiction of these two facets of the native Americans is split as well, there being no place in the military history for the sentimental reunion scenes, and no place in the reunion scenes for the military actions that made such scenes possible. The consequences of the capture of Europeans by the American Indians, particularly the acculturation of young European children to the native way of life, is a theme that Crèvecoeur would take up in his final chapter of *Letters*. One source among many for his information about such captures was a book like Smith's.

William Stork, by contrast, ignores the local American Indians completely in his *Description of East Florida*. He notes the evidence for the native Americans' past numbers: "By the best accounts of the first discovery of East-Florida, it appears to have been nearly as full of inhabitants as Peru and Mexico; and these accounts are, in some measure, verified, by the frequent remains we find of Indian towns throughout the peninsula. The natives are described to have been larger, and of a stronger make than the Mexico Indians."[67] This is the single reference in the entire book to native Americans. He does not mention the Creeks and the Seminoles in his topically arranged survey of East Florida's commercial possibilities. Bartram, who may have known Stork's book, writes similar passages in which he describes the town sites of the ancient inhabitants of Florida but makes no systematic effort to connect these people to the present American Indian inhabitants.

James Adair's *History of the American Indians* begins with twenty-three chapter-length arguments demonstrating that the natives are actually descendants of the lost tribes of Israel. The history referred to in Adair's title is confined to the American Indians' origin. Two final sections form a separate manners-and-customs account that collects in an objective description the observations he made during his forty-year residence among native Americans. The manners-and-customs portrait of the appearance, customs, manners, dress, religion, and other common characteristics of a given people is an old, stable subgenre occurring as early as Mandeville's *Travels,* a fourteenth-century text, and persisting into the nineteenth century.[68] Except for Rogers's *Journals,* each of the six literary forebears of Bartram, Jefferson, and Crèvecoeur devotes a separate section to a manners-and-customs description of the people they viewed. These representations formed a distinct countertradition to what Adair called "romance and a mass of fiction."[69]

The manners-and-customs portraits compose the portion of Adair's book that has endured. They are the most recognizably scientific in tone and style: "Ball-playing is their chief and most favourite game. . . . The ball is made of a piece of scraped deer-skin, moistened, and stuffed hard with deer's hair, and strongly sewed with deer's sinews.—The ball-sticks are about two feet long, the lower end somewhat resembling the palm of a hand, and which are worked with deer-skin thongs."[70] He goes on to describe the game of lacrosse using the same discourse as that used to describe their agriculture, feasts, and clothing. In Adair's representation of them, the American Indians have an origin, which is still a subject for conjecture, and a present existence, described in the manners-and-customs account. The time in between is simply a blank; they do not have a collective history. And unlike Smith's depiction of the American Indians, Adair's account does not include the possibility of personal histories.

Manners-and-customs accounts stand beside various narrative, including fictional, depictions of the American Indians. "Romances" in the form of captivity narratives were enormously popular in the colonies.[71] Richard Slotkin notes "the power of the captivity narrative to express the community's sense of the meaning of its experience, to rationalize its actions, and to move its people to new actions." Roy Harvey Pearce finds various colonial commentators comparing American Indians to classical heroes, devising the myth of the noble savage. Louise K. Barnett observes that American novelists, when they finally emerged, would depict the American Indian in "lifeless and narrowly conceived stereotypes," one of which is the "bad Indian," a bloodthirsty, often cannibalistic, ruthless torturer and murderer.[72] Yet critics overlook the nonfictional manners-and-customs descriptions. Pearce, for instance, reads William Bartram's "Elysian fields" encounter with American Indian women (*Travels* 289–90) but is silent about the long manners-and-customs section that forms part 4 of Bartram's text.[73] Manners-and-customs representations of the American Indian are less obvious in their stereotyping but more insidious in their consequences because the approach that produced them was considered scientific, and the descriptions themselves as objective fact. Adair's book is just one place where Bartram, Jefferson, and Crèvecoeur may have seen manners-and-customs accounts. The form, which cast the American Indians it depicted into the timeless "now" of the historical present, blinded its practitioners to the history of the people they were depicting. Like Adair, Bartram, Jefferson, and Crèvecoeur were blind to the history of the native American. In the eyes of the eighteenth century, a people without writing could not have a history.

Carver's *Travels* is half narrative and half description. The narrative account of the author's search for a northwest passage, including descriptions of the land he traveled through, precedes a long descriptive manners-and-customs section on the American Indians who lived in those territories. In one chapter, devoted to the languages he heard, he lists an English–American Indian vocabulary of the Chippewa and the Nadowessie languages.[74] Jefferson also collected American Indian vocabularies. In *Notes* he expresses the hope that such lists might ultimately solve the mystery of American Indian origins.

The diversity of themes, discourses, and aims in the representation of the American Indian provides Bartram, Jefferson, and Crèvecoeur with a tradition of established ways to create their own representations of America's aboriginal peoples. The later writers repeat the omissions of their literary forebears. They, too, fail to represent the American Indian as having a collective or individual history. They fail, for the most part, to represent him speaking. Instead they list his vocabulary. The idea that the American Indian had an existence apart from European incursion, with its ensuing battles, or from European observation, with its resultant manners-and-customs descriptions, simply was unavailable to European writers. Rogers, Smith, Stork, Adair, and Carver did not include such depictions of the American Indian. Bartram, Jefferson, and Crèvecoeur, writing after the ascendancy of natural history superimposed its own method and rhetoric on a textual tradition that was heavily weighted in favor of description, were, if anything, even less likely to represent the native Americans as having cultural or individual history.

All of these elements—the literature of place as the earlier generation left it, with its narrative and descriptive traditions; Linnaean nomenclature and the world order it implied; America's Revolution, with all that it meant for the new country—come together in the works of Bartram, Jefferson, and Crèvecoeur to form texts that define, in a scientific sense, the lands they describe. The Linnaean revolution would provide them with an intellectual framework for managing the exotic natural productions of the new land, for fitting them into an overarching order. When combined with the textual traditions of the literature of place, the Linnaean method, with its distinctive diction and rhetoric, provided Bartram, Jefferson, and Crèvecoeur with a way to depict the new in the context of the known, a way to devise a text that could verbally encompass the land.

• CHAPTER TWO •

Description and Narration in Bartram's Travels

Courtesy of the Ewell Sale Stewart Library, the Academy of Natural Sciences of Philadelphia.

"Mico Chlucco the Long Warrior, or King of the Seminoles."

Drawing by William Bartram, for the frontispiece of Travels *(1791).*

As an instance of the literature of place, William Bartram's *Travels* represents large portions of the territories of North and South Carolina, Georgia, and Florida to readers eager for images of the New World they had never seen. Using the rhetoric and method of natural history, Bartram details "the furniture of the earth" to be found in these regions—the minerals and animals and, in particular, the plants. Using Edmund Burke's theory of the sublime and the beautiful, he describes the scenes through which he sailed, paddled, rode, and walked during his three-and-a-half-year journey through the Southeast. The two methods, natural history and the sublime, complement each other. Each compels notice of a different selection of the creation. The natural historical practitioner described individual items. The Burkean practitioner described entire scenes. For Bartram, both methods were objective. The natural historical method, as we have seen, relied on observation conducted according to exact procedures. Burke's theory, relying as it did on the observer's accurate reporting of his emotional responses, provided Bartram with a scientific way of representing his reactions to the scenes he saw.

The frame for both kinds of description is narrative, the defining characteristic of the travel genre. Travel books include interruptions of the forward progress of narrative to accommodate extended descriptions of countryside or city. Narrative and description are counterpoints to each other, narrative propelling the narrator forward through time and space, description halting him to detail the scene before him.

Narrative is for Bartram a potentially mediating form of discourse between the two modes of description that he employs. He describes both the things of the external world, represented in *Travels* through natural historical description, and his internal responses to the external world, represented through Burkean description of the land and the sea. Individual action, represented in *Travels* through narrative, is both external, as Bartram moves through the world, and internal, as he experiences his own action. Bartram reports his movements as he travels, but this narrative remains a mere frame for the description. It fails to provide a middle ground between the impersonal facts of natural history and the psychologically immediate sensations of Burkean aesthetics.

This failure of narrative in *Travels* is most clearly illustrated by Bartram's representations of the American Indians. When narrative fails him, he ultimately resorts to natural historical description. Natural history, the means to knowledge of a place and the provider of a universal frame for the contents of that place, is, in Bartram's hands, inadequate when it is turned upon native people. Bartram's own response to the native Americans is often given as fear or awe, the two establishing emotions of the sublime, and this leads him into perorations on the American Indians' heroic, "noble savage" natures. A representational mediate ground between the discrete facts of the manners-and-customs account and the soaring moralizing of the Burkean description is available to Bartram. Narrative would permit Bartram's reader to see American Indians in action, and might vouchsafe him and his readers a glimpse into an individual native's experience. Bartram makes a few tentative narrative forays in the first three parts of *Travels*. Then, in part 4, he resorts to a manners-and-customs description as the rhetoric and method of natural history provide him with the most available means to represent America's original inhabitants.

Bartram was born in 1739 near Philadelphia, where his father, one of the first botanists in America, had founded a botanical garden on the banks of the Schuylkill. He traveled from 1773 to 1778 through the American Southeast under the patronage of Dr. John Fothergill of London, returned to his family's garden, and wrote an account of his journey, *Travels through North & South Carolina, Georgia, East & West Florida, The Cherokee Country, the Extensive Territories of the Muscogulges, or Creek Confederacy, and the Country of the Chactaws; containing An Account of the Soil and Natural Productions of those Regions, together with*

Observations on the Manners of the Indians.[1] During Bartram's lifetime, *Travels* was published in Philadelphia (1791), London (1791), Dublin (1793), Berlin (1793), Haarlem (1794–97), and Paris (1799).[2] This book, Bartram's only major publication, found an audience in England and on the Continent, eventually entering the canon of American literature through the auspices of readers such as Wordsworth, Coleridge, Emerson, and Thoreau.

Commentators on *Travels* have established a tradition of ignoring the natural history in the book. The early literary- source hunters did not have to concern themselves with Bartram's scientific accomplishments. Lane Cooper and later John Livingston Lowes simply followed the echoes of Bartram's text in the poetry of Wordsworth and Coleridge. Although N. Bryllion Fagin included in his book a consideration of Bartram's life, philosophy (including his science), and landscape, he too concentrated on an assessment of Bartram's influence on other writers.[3]

More recently, Robert Arner and Wayne Franklin have examined only the narrative portions of the text (parts 1 through 3) to discover the pastoral and chivalric elements of *Travels*. Thomas Vance Barnett extends this appropriation of the text to narrowly literary uses. He declares, after a demonstration that William Bartram was "a mediocre scientist" that, in fact, "William Bartram's *Travels* is . . . a literary work."[4] Efforts to limit the text before interpreting it bespeak these critics' discomfort in dealing with science, or with those elements of a text that are scientific.

Critics may simply be following Bartram's contemporary reputation as a "difficult" son to his father, a judgment that is first recorded in the letters the patron Fothergill wrote to John Bartram about his most gifted child.[5] William Bartram's relationship with Fothergill determined his relationship with the entire natural history community. That is, it determined his pre-*Travels* reputation.

William Bartram was first brought to the notice of Dr. John Fothergill (1712–80) by Peter Collinson, who showed Bartram's drawings of butterflies and plants to many members of the natural history circle.[6] Fothergill was a physician, and owner of the largest private garden in England. Adjoining a sitting room at his Upton home was a 260-foot greenhouse.[7] He owned more than 3,000 different species of exotic plants.[8] He prized botanical illustrations, employing artists to draw new items as they were added to his collection. In 1770 he agreed to buy any drawing that Bartram might send; in 1773, he agreed to sponsor Bartram on a trip to the Southeast.[9]

As we have seen in chapter 1, patrons could be self- interested purchasers of a service. Fothergill, like Collinson, was not motivated solely by the pure spirit of science. For patrons of plant hunters, their gardens "proved to be the most conspicuous means of enjoying the natural riches of the far corners of the earth." Much is often made of Fothergill's generosity in thinking of and supporting William Bartram.[10] Less is made of the advantages he reaped—the possession of curiosities from the New World to show his botanizing friends and to help him maintain his status in the British collectors' community.

In his letter to John offering to pay William's expenses on his trip to the Southeast, Fothergill made one thing clear: "I would not have it understood that I mean to support him." Indeed, Fothergill was not happy about Bartram's destination:

> He proposes to go to Florida. It is a country abounding with great variety of plants, and many of them unknown. To search for these, will be of use to science in general; but I am a little selfish. I wish to introduce into this country the more hardy American plants, such as will bear our winters without much shelter. However, I shall endeavour to assist his inclination for a tour through Florida; and if he succeeds, shall, perhaps, wish him to see the back parts of Canada.[11]

Two years later Fothergill was complaining about William to John:

> I have received from him about one hundred dried specimens of plants, and some of them very curious; a very few drawings, but neither a seed nor a plant.

> I am sensible of the difficulty he is at in travelling through those inhospitable countries; but I think he should have sent me some few things as he went along. I have paid the bills he drew upon me; but must be greatly out of pocket, if he does not take some opportunity of doing what I expressly directed, which was, to send me seeds or roots of such plants, as wither by their beauty, fragrance, or other properties, might claim attention.[12]

A dried specimen is what a botanist puts in an herbarium, a library of plants. When an unknown specimen comes to light, the botanist compares it to the specimens that are already in the herbarium to determine the new plant's classification. A seed or root, once planted and growing in the possessor's garden, can always produce cuttings, blossoms, and fruit to dry and mount for an herbarium specimen. Thus, seeds and roots yield more glory for the collector, and a better return on his investment. This acquisitiveness is reflected in Fothergill's charge to William when he agreed to sponsor a plant-gathering trip: "I am not so far a systematic botanist, as to wish to have in my garden all the grasses, or other less observable, humble plants, that nature produces. The useful, the beautiful, the singular, or the fragrant, are to us the most material."[13]

When Fothergill complains that William has not sent him seeds or roots, only dried plants and drawings, he is neatly reflecting the division between the pure hunt for knowledge about plants and his desire to have the best garden he can manage. Clearly, he thought of Bartram primarily as a plant hunter. Despite this limiting characterization, Fothergill did carry Bartram's name into the natural history circle both in England and on the Continent. When Fothergill died in 1780, three years after Bartram's return from the Southeast, the American botanist was deprived of this access. Bartram had probably begun writing *Travels* by then.[14] But the war and his patron's death must have made its completion more urgent. At the time it was his single remaining opportunity for receiving the recognition he knew he deserved.

Despite his subordinate relationship to Fothergill, Bartram

transcended the role of plant hunter to become a respected and accomplished botanist. He never took a university degree, but studied botany and natural history with an acknowledged master—his father, John.[15] He was paid for his work as a plant collector.[16] He had read the revolutionizing work of Linnaeus and was aware, as we shall see, of the current concerns of an eighteenth-century botanist.[17] Bartram appears in contemporary lists of scientists. One F. A. A. Meyer, commentator on *Travels* in its German edition, included Bartram in a 1794 list of all living zoologists; Bartram was the only American thus honored.[18] In the preface to the first American botanical publication, Henry Muhlenberg lists Bartram simply as "William Bartram, Botanist."[19] The University of Pennsylvania offered him a professorship in botany, but he refused it, citing his poor health.[20] Despite Bartram's lack of a formal post, students eagerly sought him out, then went on to illustrious careers of their own, most notably Alexander Wilson, author of *The American Ornithology*.[21] One account of Bartram's influence on botany links him with thirty-three investigators and fourteen texts.[22] His scientific achievements were recognized by his election to the American Philosophical Society.[23]

A cursory glance at *Travels* itself shows Bartram to be a natural historian with a wide range of interests—geology; botany; zoology in all of its forms, including entomology, ornithology, herpetology, and malacology; as well as anthropology. Botany was his strongest interest and the area of most of his lasting contributions. Had his drawings and verbal descriptions been published promptly, he would have claimed credit for the discovery of twenty-three American species. He was an illustrator of great talent, and his work has been favorably compared to that of Georg Ehret, the leading contemporary botanical illustrator on the Continent.[24]

The methods of eighteenth-century natural history little resemble those of natural history's modern-day descendants, the life sciences. Thomas S. Kuhn reminds us of the "unhistorical stereotype drawn from science texts" that mistakenly urges us to hold early practitioners of a given science to a modern standard of what constitutes doing that particular science.[25] One popular definition holds that an enterprise is a science if information that counts as knowledge in that field is discovered through duplicatable experiments. By this criterion, natural history is not a science at all. "Scientist" is not even a term that Bartram would have applied to himself. It was coined in 1840. "Scientific," employed in the modern sense, was not widespread until after the turn of the nineteenth century. "Science" had a long history predating Bartram, but at the time he wrote it was still being used to mean any knowledge of natural phenomena. Although Bartram did not perform experiments, he would have known about experimentation—his father was one of the first botanists in America to experiment with "mule" (hybrid) plants.[26] Bartram did, however, make observations of the most painstaking sort. He classified the specimens he observed using the Linnaean system, and he communicated these results to others in his scholarly community.

If the lack of experiments makes Bartam's science hard to recognize, so does the nature of his arguments in the introduction to *Travels,* the primary repository of his theoretical pronouncements on botany. In the introduction he writes a long essay on natural history that is organized hierarchically, covering first the vegetable and then the animal kingdom. The resulting overview establishes his metaphysical stance, and his scientific one as well. He invokes ideas of design and plenitude in the form of the Great Chain. He introduces the opposing activities of contemplation and use, opposites that remain unreconciled in the entire book as they become translated into description (which, like contemplation, is passive) and narrative (which by its nature is active, like use). To an eye alert to the signs, Bartram's introduction is a catalogue of the concerns of an eighteenth-century botanist. Such a catalogue reveals the still-close association of botany and teleology, and teleology of a grand sort—the use of design or purpose to explain events in nature.

The scope of the introduction is nothing less than the entire animate world. Bartram divides it into distinct sections whose subjects are located on progressively higher links of the Great Chain of Being: an eleven-paragraph section on the vegetable world with teleological as well as physiological speculations on several of the problems then current in botany; an eight-paragraph section on the animals with anecdotes carefully chosen to illustrate their moral system; and a four-paragraph section on American Indians that constitutes a plea for considering them to be more civilized than the usual European believed them. The general movement is toward elevation—the plants Bartram mentions are animal-like; the animals are humanlike; the savages are not savage at all. For Bartram, all creatures, plants, and inanimate things yearn upward; for those who can understand its hidden reality, everything in creation has more to recommend it than was commonly thought. Bartram reads the universe as more exalted, more able, more accomplished than it was usually seen to be. In addition to this improvement plan for the universe, he uses a mode of thought characteristic of eighteenth-century botany: analogy.

The first two sentences embody the hierarchical thinking with which Bartram's prose is suffused:

> The attention of the traveller should be particularly turned, in the first place, to the various works of Nature, to mark the distinctions of the climates he may explore, and to offer such useful observations on the different productions as may occur. Men and manners undoubtedly hold the first rank— whatever may contribute to our existence is also of equal importance, whether it be found in the animal or vegetable kingdom; neither are the various articles, which tend to promote the happiness and convenience of mankind, to be disregarded.[27]

Here Bartram states his immediate end in writing: use. (Ultimately, his end was to glorify God.) "Men and manners" are at the top of the chain, then that part of creation lesser than man but living—animals and plants—then inanimate creation (the last-mentioned "various articles") at the bottom of the chain. Lovejoy reminds us of the common place that this hierarchy occupied in Bartram's era: "It was in the eighteenth century that the conception of the universe as a Chain of Being, and the principles which underlay this conception—plenitude, continuity, gradation—attained their widest diffusion and acceptance."[28] Bartram authorizes the enterprise to his British audience first by mentioning his father, botanist to King George III; then his Father: "This world, as a glorious apartment of the boundless palace of the sovereign Creator, is furnished with an infinite variety of animated scenes, inexpressibly beautiful and pleasing, equally free to the inspection and enjoyment of all his creatures."[29] This invocation of the Creator, with its reference to plenitude ("infinite variety"), at once authorizes the enterprise and determines its ultimate end. In addition to use, Bartram states a more immediate end for his enterprise—inspection and enjoyment. Throughout *Travels* Bartram claims his reason for writing is to be of use, but the text represents and invites "inspection and enjoyment," as does his practice in the field.

The second section, which begins the long discourse on the vegetables, continues this teleological bent: "There is not any part of creation, within reach of our observations, which exhibits a more glorious display of the Almighty hand, than the vegetable world." Having once again framed his remarks with mention of the universe and its creator, Bartram narrows his scope, confining himself merely to the earth: "It is difficult to pronounce which division of the earth, between the polar circles, produces greatest variety." He provides a Linnaean catalogue of tropical plants that seem intended for "luxurious scenes of splendour."[30] The temperate zone "exhibits scenes of infinitely greater variety, magnificence, and consequence, with respect to human economy, in regard to the various uses of vegetables."[31] He follows with another long list of plants of the temperate zone, which provides evidence of plenitude, one of the principles of the Great Chain of Being. The fulsome list evokes the brimming creation.

This section has another aim as well. Bartram has divided the globe into zones—a common practice today. But traveling in the colonies in 1749, Peter Kalm, a Linnaean disciple, records in his diary this conversation with William's father: "[John Bartram] reiterated what he had often told me before, namely that all plants and trees have a special latitude where they thrive best, and that the further they grow from this region, whether to the north or south, the smaller and more delicate they become, until finally they disappear entirely."[32] In 1749 this observation was noteworthy. As a topic of conversation it recurred in Kalm's talks with John Bartram. It was not simply a settled matter. The rest of their conversation that day involved instances of plants at the extreme limits of what we now call their range, including an aloe growing in Virginia.

In a section of the introduction devoted to plants that seem to straddle the boundary between plants and animals, William Bartram discusses the function of parts of the pitcher plants (*Sarracenia*), of the apparently volitional motion of both the Venus flytrap (*Dionea muscipula*), and of the tendrils of certain climbing plants, such as the cucumber (*Curcurbita*). He speculates on the little "lid" of the pitcher plant, the "cordated appendage," which nature has provided to guard against the vessel's filling with water and breaking. The plant's leaves, shaped like pitchers, cannot support a "sudden and copious supply of water from heavy showers of rain, which would bend down the leaves, never to rise again because their straight parallel nerves which extend and support them are so rigid and fragile, the leaf would inevitably break when bent down to a right angle."[33] This is our first evidence of Bartram's link to branches of natural history other than taxonomy. Bartram is speculating as to the function of certain structures he observed and sketched.[34] The prominent "nerve"—the supporting rib and its branches—is featured in each of several renderings he made of the plant. The mode of discourse here, rather than the catalogue, is analysis; the mechanism of explanation, the analogy. Parts of plants are compared to parts of animals.

In employing analogies, Bartram weaves into his work one of the dominant threads of eighteenth-century botanical discourse. For many historians of botany, the analogic method of investigation marks Bartram as, at the very least, one who would be left by the wayside when the histories of botany were finally written. In one such history Philip C. Ritterbush has recovered the place that analogy held in eighteenth-century botanical reasoning. Naturalists believed, and based arguments on the proposition, that "plants were analogous to animals because of their close proximity in the scheme of nature." His disapproval of such analogizing is plain: "Although the authority of science was invoked on [its] behalf the concept reflected an improper understanding of organic nature, far exceeded the evidence given for [it], and too often led naturalists to neglect observations and experiment in favor of abstract conceptions."[35]

A second historian of botany, François Delaporte, sets himself against historians like Ritterbush who vilify the analogists and champion the experimentalists:

> The analogical method was a hindrance, it is argued, because it was a procedure the eighteenth century presumably inherited from the Renaissance or even Antiquity. The identification of the various parts of the vegetable organism with the known parts of the animal was, we are told, a source of error. In contrast, the work done by experimentalists and observers was supposedly a prefiguration of the nineteenth century.

This separation of the botanists into two camps—the successful observers and the misguided analogists—is problematic, says Delaporte.

> That the discipline has had to be divided up in this way is one indication that something has gone wrong. For one thing, the logical connections among certain statements are obscured, because botanists numbered among the analogists happen to have become involved in observation or experiment. For another, the use of analogy by those described as observers and experimentalists is disguised.[36]

Ritterbush discovers the relationship between the Great Chain, the usual eighteenth-century way of visualizing the

Creator's plan, and analogy: "There was a reciprocal relationship between analogies and imputed proximity of place on the scale. The discovery of certain properties which plants shared with animals lent encouragement to the belief that they were close to animals upon the scale of beings and that consequently any number of analogies might pertain between them."[37] Analysis—what Bartram calls *analysis* moderns would likely call *observation and description*—went hand in hand with reasoning from analogy. And since there was a plan, one could reason not only from the Creation to the blueprint for it, but from the blueprint back to the Creation. This is much more consequential than the accidentally heuristic models that twentieth-century scientists construct. This same correspondence, this same path to truth, was also possible with the parts of the Creation—plants were believed to be like animals because they shared with them certain functions such as nourishing themselves, reproducing themselves, and moving volitionally. One could reason from the better-known animals to the lesser-known plants just as one could reason from the blueprint to the Creation. Analogy was not simply a way to explain a lesser-known thing in terms of a better-known thing—a one-way flow of knowledge; instead, it was a way to know both things in terms of each other—a two-way flow.

This speculative reasoning in the introduction is quite different from the taxonomy that is the focus of parts 1–4 of Bartram's *Travels*. It places Bartram's thought in a broader context. The Great Chain of Being figures not only as the overarching scheme for the taxonomic work that Bartram did but also as a sort of metaphysical abacus. A plant's essence, like a counter on an abacus rod, could be moved higher, toward the animals, or lower, toward inanimate matter, through this analogical thinking. By choosing certain difficult cases, such as an animal-like "carnivorous vegetable" (the Venus flytrap) and arguing from observable characteristics to metaphysical states, Bartram could slide enough counters toward a higher state of being to shift the whole nature of the Creation. Analogy between plants and animals had led Linnaeus to his "sexual system" of taxonomy. (Reproduction in animals was understood before reproduction in plants.) Bartram, in his speculations here, was in tune with the scientific methods of his day, however odd it now seems to us that he would try to demonstrate the animality of certain plants.

This elevation of the things in the "glorious apartment of the Creator" is carried through to his observations of certain animals. Bears seem humanlike in their "parental and filial affections."[38] Spiders are "cunning" and "intrepid" in their hunting; birds, "social and benevolent creatures."[39] But if one consequence of his analogizing is his exaltation and sentimentalization of animals, another is his willingness to look at American Indians in a spirit of equality. The final paragraphs of the introduction are Bartram's plea to European settlers to send visitors to the American Indians "to learn perfectly their languages, and by a liberal and friendly intimacy become acquainted with their customs and usages, religious and civil; their system of legislation and police, as well as their most ancient and present traditions and history." This is quite a program. Its goal might be assimilation: "[The Indians] were desirous of becoming united with us, in civil and religious society," but it is assimilation based on thorough knowledge.[40] He does not suggest a missionary excursion where European visitors would teach the native Americans an imported language, traditions, and history; rather, he suggests exactly the kind of visit that he himself makes in *Travels*—to "learn perfectly" and to return "to make true and just reports." He suggests, in short, an expedition whose methodology is scientific within the broad compass of natural history.

Bartram's introduction, written after his return from the Southeast, puts into motion the elements of his science and of his thinking, and offers us a way to understand where many of his more-commented-upon ideas came from. The universe is a hierarchy, traditionally divided into three kingdoms—minerals, plants, and animals—with man at the top of animal creation. But science shows Bartram a way of blurring these categories, or employing them to redefine them: plants are like animals, animals are like men, and American Indians are like Europeans. More particularly, certain characteristics that we ascribe to animals, such as volitional movement, are also present in plants, and this blurring of the boundaries between plants and animals is grounds for reevaluating creation wholesale—and for promoting its constituents upward in our estimation of them. So animals display humanlike characteristics, and since in humans the source of these characteristics is a refined and reasoning intellect, animals must have one, too. So, too, do American Indians display European-like characteristics. They must also have similar sources for such beliefs and will benefit from the same sort of treatment that Europeans might expect under similar circumstances.

Within this universal teleology is a program for man's relationship with the natural world. There are two choices: observation and use. In Bartram's text, the corollary of observation is description; the corollary of use, narration. For Bartram traveling through the Southeast, observation and use define his possibilities for interacting with the territories through which he passes. For Bartram writing *Travels,* the two associated modes of discourse define the possibilities for representing that country in the text. *Travels* contains both modes: description on the Linnaean and Burkean models, and narrative.

Bartram was trained in the Linnaean system. His father had received a copy of Linnaeus's *Systema Naturae* in 1736, three years before William's birth, and had been tutored in its use by James Logan, an accomplished colonial botanist who demonstrated, experimentally, the mechanism by which pollen fertilized corn.[41] Bartram's mastery of the Linnaean system is best demonstrated by his account of his most famous discovery, *Franklinia alatamaha.*

> It is a flowering tree, of the first order for beauty and fragrance of blossoms: the tree grows fifteen or twenty feet high, branching alternately; the leaves are oblong, broadest towards their extremities, and terminate with an acute point, which is generally a little reflexed; they are lightly serrated, attenuate downwards,

> and sessile, or have very short petioles; they are placed in alternate order, and towards the extremities of the twigs are very large, expand themselves perfectly, are of a snow white colour, and ornamented with a crown or tassel of gold coloured refulgent staminae in the centre, the inferiour petal or segment of the corolla is hollow, formed like a cap or helmet, and entirely includes the other four, until the moment of expansion; its exterior surface is covered with a short silky hair; the borders of the petals are sessile in the bosom of the leaves, and being near together towards the extremities of the twigs, and usually many expanded at the same time, make a gay appearance: the fruit is a large, round, dry woody apple or pericarp, opening at each end oppositely by five alternate fissures, containing ten cells, each replete with dry woody cuneiform seed.[42]

The concentration on the flower and the fruit is the hallmark of Linnaeus's sexual system of classification.[43] Other characteristics are not ignored, but these are the most closely scrutinized. Hence Bartram's remark about his first look at the shrub: "This very curious tree was first taken notice of about ten or twelve years ago, at this place, when I attended my father (John Bartram) on a botanical excursion; but it being then late in the autumn, we could form no opinion to what class or tribe it belonged." *Franklinia* blooms in the spring. He adds this footnote:

> On first observing the fructification and habit of this tree, I was inclined to believe it a species of *Gordonia;* but afterwards, upon stricter examination, and comparing its flowers and fruit with those of the *Gordonia lasianthus,* I presently found striking characteristics abundantly sufficient to separate it from that genus, and to establish it the head of a new tribe, which we have honoured with the name of the illustrious Dr. Benjamin Franklin.

Bartram was obviously more than a simple Quaker boy who rode through the American Southeast gathering new-looking plants and sending them back to England. The comparison of the flowers and fruit of one plant with those of another—the most essential procedure in naming a plant using the Linnaean method of classification—also marks him as more than a literary-minded pilgrim on a pastoral retreat.[44]

The rhetoric imposed by the Linnaean method resulted in a series of descriptions, all of them static or, as in the case of *Franklinia,* atemporal, where blossom and fruit are represented in the same description. Even as they suspend or transcend time they establish the physical reality of the country. To represent the Southeast's natural history, Bartram had at his disposal three different forms of natural historical description: an entry describing a single item, a list of like items, and an essay encompassing the range of natural history in a given locale.

In the third chapter of the first volume of *Travels,* Bartram offers the reader a description of "a new species of Anona." This is an instance of an entry describing a single item. The passage serves as an "exploded" version of an entry in a work like the *Species Plantarum.* It also serves as a verbal gloss on the engraving that appears on the facing page of the London 1792 edition.[45] The text-and-illustration combination permits the reader to experience the method of natural history—to look at the specimen and at the same time to read the description that Bartram has produced.

"It is very dwarf, the stems seldom extending from the earth more than a foot or eighteen inches." With this scale in mind the reader then learns, "The leaves are long, extremely narrow, almost lineal." A glance at the picture (reprinted here) confirms this. "However, small as they are, they retain the figure common to the species, that is, lanceolate, broadest at the upper end and attenuating down to the petiole [leaf stalk] which is very short." This is especially clear in the largest, centermost leaf. "Their leaves stand alternately, nearly erect, forming two series, or wings, on the arcuated [bow-shaped] stems." The drawing clearly shows the alternating leaves, the bowed stem. "The flowers, both in size and colour, resemble those of the Antrilove"—this was beyond his technical means to show in the drawing—"and are single from the axillae of the leaves [the upper angle between a leaf or petiole and the stem from

ANONA PYGMEA.

which it springs] on incurved pedunculi [stems bearing single flowers], nodding downwards." The uppermost flower best shows this relationship between stem and leaf, the nontechnical "nodding" mirroring the "incurved pedunculi" that produce this downward-looking attitude.

This careful description and the equally careful drawing teach the reader how to look at all of the drawings and read all of the descriptions in the volume. The details of leaf shape, attachment, and flower position are precise. The verbal description enables a nonbotanical reader to understand much of the information, particularly with the aid of the drawing. Foucault explains this close association of the verbal and the visual. He notes that "the blind man in the eighteenth century can perfectly well be a geometrician, but he cannot be a naturalist." In natural history, sight has "an almost exclusive privilege, being the sense by which we perceive extent and establish proof." But sight itself has been narrowed, refined, reduced to the most certain of its elements. It is "a visibility freed from all other sensory burdens and restricted, moreover, to black and white." Bartram's illustrations for this edition of *Travels* were black-and-white line drawings. Visibility determines natural history: "This area, much more than the receptivity and attention at last being granted to things themselves, defines natural history's condition of possibility, and the appearance of its screened objects: lines, surfaces, forms, reliefs."[46] Foucault notes that the natural historian's sight operates within four variables: "The form of the elements, the quality of those elements, the manner in which they are distributed in space in relation to each other, and the relative magnitude of each element." Variations in form, quality, distribution, and magnitude of five parts of the plant—"roots, stem, leaves, flowers, fruits"—become the keys to the plant's identification. He asserts that the relationship between the name and that which it denotes, between language and things, "can . . . be established in a manner that excludes all uncertainty."[47] Furthermore, "the plant is thus engraved in the material of the language into which it has been transposed, and recomposes its pure form before the reader's very eyes. The book becomes the herbarium of living structures."[48]

In Bartram's description of the Anona we have the apotheosis of the Linnaean form of natural historical inquiry. A single plant is reconstituted in the mind of the reader by a verbal description. Bartram's volume included seven illustrations of natural historical specimens, yet it contained more than fifty verbal descriptions of the flora and fauna of America. He relied on the shared language of natural historical description to represent the contents of the American Southeast to the reader. The specimens and the book are made one by the rhetoric of the Linnaean method. Samples of American plants and animals are included in every copy of the book.

As we saw in chapter 1, every Linnaean binomial is backed by a description similar to the one that Bartram offered of his "new species of Anona." When Bartram includes a Linnaean name in *Travels*, he invokes the entire intellectual scheme upon which the Linnaean system was built—the unique description of a single plant as well as that plant's place in the Great Chain. If the invocation of this context made the name an unambiguous representation of a part of America, it also abstracted the thing represented out of its American context. Bartram's lists compensate for this deficiency in the rhetoric that grew out of the method.

In describing an island off the coast of Sunbury, Georgia, Bartram includes lists of all of "the natural produce of these testaceous ridges"—the plants and animals.[49]

> The general surface of the island being low, and generally level, produces a very great variety of trees, shrubs, and herbaceous plants; particularly the great long-leaved Pitch-Pine, or Broom-Pine, *Pinus palustris, Pinus squamosa, Pinus lutea, Gordonia Lasianthus,* Liquid ambar *(Styraciflua) Acer rubrum, Fraxinus excelcior; Fraxinus aquatica, Quercus aquatica, Quercus phillos, Quercus dentata, Quercus humila varietas, Vaccinium varietas, Andromeda varietas, Prinos varietas, Ilex varietas, Viburnum prunifolium, V. dentatum, Cornus florida, C. alba, C. sanguinea, Carpinus betula, C. Ostrya, Itea Clethra alnifolia, Halesia tetraptera, H. diptera, Iva, Rhamnus frangula, Callicarpa, Morus rubra, Sapindus, Cassine,* and of such as grow near water-courses, round about ponds and savannas, *Fothergilla gardini, Myrica cerifera, Olea Americana, Cyrilla recemiflora, Magnolia glauca, Magnolia pyramidata, Cercis, Kalmia angustifolia, Kalmia ciliata, Chionanthus, Cephalanthos, Aesculus parva.*[50]

These names would appear in a work like Linnaeus's *Species Plantarum* or Gronovius's *Flora Virginica* separated from each other by many pages of text. *Fraxinius excelcior,* for example, appears in the first edition of the *Species Plantarum* grouped with the *Polygamia Dioecia* on page 1057. There it is listed with the other *Fraxinia,* which follow the *Gleditsia* and are followed by the *Diospyros.* It occupies its niche both in the *Species Plantarum* and in the Great Chain, with gradations on one side or another duly listed. The *Magnolias* are listed with the *Polyandria Polygynia*, on page 536. Yet in Bartram's book, the two are listed in the same paragraph at the same place in the text, and so are located for the reader on the island. They are entries in the 1200-page Linnaean table of plants, where they appear with their near neighbors on the Great Chain. They are also entries in Bartram's text, where they appear with their actual neighbors in the world. Bartram's list counters the abstracting influence of the Linnaean name. It provides a local habitation or context to accompany the universal name and location on the Chain. The list represents to the reader an American scene.

Taken together, Bartram's lists of the natural productions of Georgia's Atlantic islands represent to the reader an American territory, located (off the coast of Sunbury), bounded (the islands themselves), and furnished with the items in the lists. When he made the survey on which he based his description in *Travels*, Bartram was revisiting the islands. They merited a second look because he knew they would "exhibit a comprehensive epitome of the history of all the sea-coast Islands."[51] In his *Report* to Fothergill, prepared from his field notes and sent to England while Bartram was still traveling, he wrote for his patron a four-sentence account of the islands. On them he had "discover'd nothing new, or much worth your notice."[52] In *Travels*, the description of the islands is ten times longer. Clearly, Bartram was interested in showing the reader of *Travels* not just what was new on the islands, but whatever was there. The essay describing the natural history of an entire area represented for the reader a definitely bounded piece of American territory, a country in the geographical sense of the word.

The natural historical entry, list, and essay define space at the expense of time: they suspend narrative. Bartram's visit to the islands begins in narrative: "Next day, being desirous of visiting the islands, I forded a narrow shoal, part of the sound, and landed on one of them, which employed me the whole day to explore." Curiously, it ends in the same bit of action: "The sight of this delightful and productive island, placed in front of the rising city of Sunbury, quickly induced me to explore it; which I apprehended, from former visits to this coast, would exhibit a comprehensive epitome of the history of all the sea-coast Islands of Carolina and Georgia, as likewise in general of the coast of the main."[53] The opening statement is truly narrative, containing the details of Bartram's path from the mainland to the island. But the closing passage takes us back to a moment before the opening passage, to Bartram's seeing the island and the city of Sunbury, and to knowledge that he had from a prior visit ("former visits to this coast"). In between the parenthetical narrative statements we read about the island's soil, artifacts (fragments of earthen vessels), plants, and animals. In these passages, particularly in the lists of plants with their Linnaean names, the temporal element present in the narrative drops out, and we are presented with a list located in space, but no longer in time. As an observer moving through time, Bartram disappears. He reappears as the describer who is not in time ("throughout the seasons"). Time suspends before the calling of the eternal Linnaean names. The static, curiously still descriptions partake of the verbless nature of the Linnaean names themselves. The reader reconstitutes the scene, but does so in the historyless "now" that is a consequence of the Linnaean rhetoric.

The other descriptive method used in *Travels* is Edmund Burke's. "Romantic" passages conveying Bartram's awe and terror at the vistas he encountered have received more critical attention than the natural historical description.[54] Critics assume that aesthetics and science must be warring points of view. If a writer is romantic, he must not be scientific. If a writer is scientific, he cannot truly be said to be romantic. These divisions are modern, imposed from a perspective in which science is looked upon as positivism and art as extrarational. But in Bartram's day, and in Bartram's text, science and art are not at war with each other.

The sublime makes its appearance early in *Travels*, and is invoked throughout to convey the immensity of the objects and the magnitude of the experience that Bartram faced. Edmund

Burke is the most important eighteenth-century theorist of the sublime, and there is evidence that Bartram saw his work. A survey of treatises on art and aesthetics available in America shows among its forty-seven entries for Burke's *Philosophical Enquiry into the Origin of Our Ideas of the Sublime and Beautiful* that as early as 1760 a copy was offered for sale in a Philadelphia bookseller's catalogue. At some point before 1807 a French translation of the work was donated to the Library Company of Philadelphia, an institution where Bartram would have had privileges. And in 1771, in a letter recommending books appropriate for a gentleman's library, Thomas Jefferson included Burke.[55] Bartram lived within a morning's ride of Philadelphia, a city that certainly knew of Burke's theory and that contained copies of Burke's *Enquiry*. He had the opportunity to know Burke's book, as his own text demonstrates.

In the early pages of *Travels*, when Bartram introduces the reader to the ideas that will be needed to navigate the book itself, he includes an account of a storm that lasted for the second and third days of his voyage from Philadelphia to Charleston:

> The powerful winds, now rushing forth from their secret abodes, suddenly spread terror and devastation; and the wide ocean, which, a few moments past, was gentle and placid, is now thrown into disorder, and heaped into mountains, whose white curling crests seem to sweep the skies!
>
> This furious gale continued near two days and nights, and not a little damaged our sails, cabin furniture, and state-rooms, besides retarding our passage.[56]

Contrast this account with Bartram's interpretation of ocean travel, offered just one paragraph later:

> There are few objects out at sea to attract the notice of the traveller, but what are sublime, awful, and majestic: the seas themselves, in a tempest, exhibit a tremendous scene, where the winds assert their power, and, in furious conflict, seem to set the ocean on fire. On the other hand, nothing can be more sublime than the view

> of the encircling horizon, after the turbulent winds have taken their flight, and the lately agitated bosom of the deep has again become calm and pacific; the gentle moon rising in dignity from the east, attended by thousands of glittering orbs; the luminous appearance of the seas at night, when all the waters seem transmuted into liquid silver; the prodigious bands of porpoises foreboding tempest, that appear to cover the ocean; the mighty whale, sovereign of the watery realms, who cleaves the seas in his course; the sudden appearance of land from the sea, the strand stretching each way, beyond the utmost reach of sight; the alternate appearance and recess of the coast, whilst the far distant blue hills slowly retreat and disappear; or, as we approach the coast, the capes and promontories first strike our sight, emerging from the watery expanse, and, like mighty giants, elevating their crests towards the skies; the water suddenly alive with its scaly inhabitants; squadrons of sea-fowl sweeping through the air, impregnated with the breath of fragrant aromatic trees and flowers; the amplitude and magnificence of these scenes are great indeed, and may present to the imagination, an idea of the first appearance of the earth to man at the creation.[57]

This is not so much a description of a storm—the earlier, four-sentence account is such a description—as it is a theory of what looking at the ocean is like, of the contrast between stormy seas and quiet ones, generalized to apply to almost any ocean voyage, to any view at sea. The ocean, in other words, causes certain effects in the viewer, who is generalized along with Bartram ("we approach") until both fade into the perceiving imagination of Adam himself.

This account of the effects of the ocean on the viewer, indeed on the first viewer, is put forth in Burke's *Enquiry* as productive of the sublime: "A level plain of a vast extent on land, is certainly no mean idea; the prospect of such a plain may be as extensive as a prospect of the ocean; but can it ever fill the mind with any thing so great as the ocean itself? This is owing to several causes, but it is owing to none more than this, that the ocean is an object of no small terror."[58] Burke's textbook example of sublimity is also Bartram's. And although it appears in a book about a particular man traveling a particular route, Bartram couches his explanation in the broadest terms. This is partly because he seems self-conscious, in those opening pages, of introducing the reader to the kind of analysis he will be doing in the book itself—interested, that is, in giving the reader the essential frameworks from which he will be operating.

But Bartram's explanation goes beyond Burke's, which had been offered in the section entitled "Terror," to provide an example of the terrible producing the sublime.[59] Bartram, instead, provides us in this first instance of sublimity in *Travels* with a sort of compendium of the sublime. He adds details that support most of the elements that Burke notes as productive of the sublime. In the shorter, specific description Bartram offers us terror and devastation; then, in the more generalized lecture on the sublime, he includes the "encircling horizon" (vastness), the moon with its "thousands of glittering orbs" (vastness and intensity of light), the "luminous appearance of the seas . . . transmuted into liquid silver" (intensity of light, vastness), "prodigious bands of porpoises foreboding tempest" (the threat of danger), "the mighty whale" (magnificence), "who cleaves the seas in his course" (power), "the sudden appearance of land from the sea" (suddenness), "the alternate appearance and recess of the coast" (intermittence), "promontories . . . like mighty giants" (vastness), "the water suddenly alive" (suddenness), and finally, what seems to be Bartram's own contribution to the idea of the sublime, "the air impregnated with the breath of fragrant aromatic trees and flowers (the opposite of Burke's intolerable stenches). Before he has even gotten off of the boat that took him south, Bartram's retrospective narrator has offered the reader a compendium of the sublime. All of these elements reappear in the narrative, over and over, as Bartram beholds still another prospect whose awesome or terrible appearance compels him to convey his reactions.

Both the American traveler and the British aesthetician trace the source of these reactions to the same emotions. Bartram nominates "curiosity" as the "attendant spirit" of his travels.

Burke, in the first sentence of part 1 of the *Enquiry,* says, "The first and the simplest emotion which we discover in the human mind, is Curiosity." Bartram traveled in search of new plants for Fothergill, and new scenes to gratify his own curiosity. Burke's theory also recognizes the importance of the new: "Some degree of novelty must be one of the materials in every instrument which works upon the mind; and curiosity blends itself more or less with all our passions." The sources, then, of passion and information, for Bartram and for Burke, were the same—novelty and its motivating spirit, curiosity.[60]

Both men also shared a belief in the importance of direct observation. In defending the limitations of his theory Burke noted, "A theory founded on experiment and not assumed, is always good for so much as it explains."[61] Burke's departure from previous aesthetic practice was to examine his own reactions rather than to simply convey the findings of the rhetoricians who had preceded him.[62] Bartram's entire method was one of seeing and observing. The very enterprise of traveling to observe attests to that. Indeed, this similarity in aims—from the beginning, Bartram speaks of "observations" and the importance of the visual (Burke's theory is a visual one)—is everywhere apparent in *Travels*. Burke's theory is also an unusually democratic one: "The true standard of the arts is in every man's power; and an easy observation of the most common, sometimes of the meanest things in nature, will give the truest lights, where the greatest sagacity and industry that slights such observation, must leave us in the dark, or what is worse, amuse and mislead us by false lights."[63]

Yet Burke is not claiming that each observer makes up his own mind independent of other observers. Having made the reactions of one person the basis of his theory, he goes on to assert that all men have similar reactions—"the standard both of reason and Taste is the same in all human creatures." He offers a definition of *taste:* "I mean by the word Taste no more than that faculty, or those faculties of the mind which are affected with, or which form a judgment of the works of imagination and the elegant arts." And he makes it even stronger:

"The principle of pleasure derived from sight is the same in all."[64] Thus, Bartram, subscribing to Burke's theory, could report the effects of the vistas on the viewer with the same sort of belief in them that he held in his observations of plants and animals. Burke was scientific in the sense that those reports of rhapsody were the same, for Bartram, as reports of a new plant or animal. They were observations that were privileged, because they depended upon faculties rather than upon changeable judgment. Bartram's description of the ocean is akin to his description of the island off Sunbury, offering the reader an epitome of the sublime in much the same way that a description of the island offered the reader an epitome of the natural products of the seacoast islands.

Bartram describes a storm that is in time, then describes what oceans do to any viewer at any time. This analysis is presented as eternal, as embodying the same sort of principles that Adam himself operated under at "the first appearance of the earth to man." The universal behind the particular is Bartram's aim, reducing experience to a set of responses that the responder can count on simply because he is in possession of the same set of equipment as other responders.

Report to Fothergill, the field journal Bartram kept for his patron, does not include the rhapsodic reactions to the scenery in which *Travels* abounds. It is an accounting of the natural history that Bartram saw on his way, composed with an eye to Fothergill's already considerable knowledge of American animals and, particularly, plants. Bartram wrote the *Travels,* including the reactions to sublime scenes, from the garden on the Schuylkill, within an hour's ride of Philadelphia. How could he have done this?

Trained as a painter, Bartram must have had a good visual memory. And the field journals contained enough detail to remind him of what he had seen and when—to a point. The rest was a matter of considering the objects he had encountered—the stimulus—and applying the appropriate term of sublimity or beauty—delineating the response. This somewhat mechanical procedure accounts for the repetitive, formulaic effusions over yet another sublime forest, yet another vast savanna.[65] Understanding the mechanism, however, need not lessen the impact of the descriptions. Bartram was showing his readers these places for the first time. Clearly, Bartram's rhapsody did not somehow consume his science. It was not founded differently from his science. Epistemologically, the book is of a piece. Aesthetics and science coexist in a telling based on experience filtered through Linnaeus's system or through Burke's, both of which are built on the bedrock of reporting what an observer experienced.

Any accounting of Bartram's aesthetics would be incomplete without a consideration of the other key term in eighteenth-century aesthetics: the picturesque. S. H. Monk notes "the invention, late in the century, of a third category, the picturesque, which had to come into existence in order to give those objects that are neither beautiful nor sublime (in Burke's sense of the words) a local habitation and a name."[66] In his treatment of the picturesque, Bartram rejects an aesthetics based on something other than experience or observation.

The picturesque was known to Bartram—he uses the term carefully. Christopher Hussey defines it as the "habit of viewing and criticizing nature as if it were an infinite series of more or less well composed subjects for painting."[67] Traveling was an important way to encounter the picturesque; travel accounts were published describing scenes in picturesque fashion and were, in turn, used as guidebooks to teach a novice traveler where to find such scenes. "The picturesque traveller is the traveller who has a conception of an ideal form of nature, derived from landscape painting, and whose purpose it is to discover ideal scenes in existence."[68] The composition of the scene was prearranged. The traveler simply went in search of it, confirming, when he found it, his sense of beauty, derived from painting.

Bartram rejects this intercession of art between observer and object. In a famous passage in the second part of *Travels*, Bartram describes the "paradise of fish" in a sinkhole filled by a subterranean spring. In the resulting deep well of absolutely clear water the fish swim "with free and unsuspicious intercourse." The basin is remarkable. The water is so clear, the fish so unafraid of one another, their disappearance into and reappearance from a sunken corridor so impressive, the scene excites "terror and astonishment."[69]

The basin is so very remarkable that Bartram takes pains to remove himself from charges of ornamenting his description:

> This amazing and delightful scene, though real, appears at first but as a piece of excellent painting; there seems no medium; you imagine the picture to be within a few inches of your eyes, and that you may without the least difficulty touch any one of the fish, or put your finger on the crocodile's eye, when it really is twenty or thirty feet under water.
>
> And although this paradise of fish may seem to exhibit a just representation of the peaceable and happy state of nature which existed before the fall, yet in reality it is a mere representation; for the nature of the fish is the same as if they were in Lake George or the river.[70]

The prelapsarian ideal society demonstrated by the fish in this basin, which seemed at first to be a mediumless painting, is a "mere" not a "just representation," and the reader is to realize that the fish's true natures have not been idealized from their lake and river roles of prey and predator. Neither has the description of the basin been idealized; the behavior of the fish can be accounted for by the impossibility of ambush in the clear water, thus depriving predators of necessary covert from which to take their prey, covert that they have in Lake George and the river. The delusive element here, for Bartram, is not his description of the scene, but the scene itself. He recognizes in it the sort of idealized landscape that artists might paint, and he cautions the reader to reject this model. He offers instead a demythologized interpretation that uses the nature of the fish to account for their behavior in this seemingly mediumless medium that tempts the describer to reach for the marvelous.

Elsewhere "mere representation," where it approaches the

ideal of the picturesque, is also subject for suspicion. Bartram is describing the Sand Hills, which have

> the appearance of the mountainous swell of the ocean immediately after a tempest; but as we approach them, they insensibly disappear, and seem to be lost; and we should be ready to conclude all to be a visionary scene, were it not for the sparkling ponds and lakes, which at the same time gleam through the open forests, before us and on every side, retaining them in the eye, until we come up with them. And at last the imagination remains flattered and dubious, by their uniformity, being mostly circular or elliptical, and almost surrounded with expansive green meadows; and always a picturesque dark grove of live oak, magnolia, gordonia, and the fragrant orange, encircling a rocky shaded grotto of transparent water, on some border of the pond or lake; which, without the aid of any poetic fable, one might naturally suppose to be the sacred abode or temporary residence of the guardian spirit.

This is the account of a prospect that is too ideal, a reality too much like paintings of nature. Bartram punctures the illusion with his last phrase, "but is actually the possession and retreat of a thundering absolute crocodile."[71] Bartram mistrusted the picturesque ideal.

In another passage Bartram completes a description with picturesque details, while at the same time labeling the completion as delusive. He is describing water lettuce, *Pistia stratiotes*. He offers a complete botanical description, then notes,

> These floating islands present a very entertaining prospect: for although we behold an assemblage of the primary productions of nature only, yet the imagination seems to remain in suspense and doubt; as in order to enliven the delusion and form a most picturesque appearance, we see not only flowery plants, clumps of shrubs, old weather-beaten trees, hoary and barbed, with the long moss waving from their snags, but we also see them completely inhabited, and alive, with crocodiles, serpents, frogs, otters, crows, herons, curlews, jackdaws, &c. There seems, in short, nothing wanted but the appearance of a wigwam and a canoe to complete the scene.[72]

The completion of the scene is a picturesque one—and for Bartram this meant the imposition on nature of a visual ideal that one carried into the wilderness rather than the representation of a real wilderness when one had gone forth to discover her novelties.

The Burkean mode of description provides Bartram with a means of objectively representing his reactions to the locales through which he travels. Taken together with the natural historical mode, the two forms of description represent to the reader the natural scenes observed in America and the effects of those scenes on the observer. *Travels* is ruled by the rhetoric of these modes of description, a circumstance that becomes problematical when Bartram uses these modes to describe the highest link of the Great Chain of Being—the American Indians. In the narrative portion of *Travels*, parts 1 through 3, Bartram shows some awareness of the problems of writing narrative about native people, but ultimately his literary models for such narrative prove inadequate, and his natural historical models assert themselves to provide the reader with the most extended passage about the American Indians in *Travels*—part 4.

As we have seen, Bartram arrives at a statement of philosophy concerning the native Americans in his introduction to *Travels*. He proposes that the Indians, who are "desirous of becoming united with" European society, be made the subject of extensive study by emissaries from European culture before newcomers to the American Indians' land formulate any plan concerning them. That is, he proposes what would now be called a field study, in which proto-anthropologists would live in "liberal and friendly intimacy" in American Indian towns, learning the language and customs of the inhabitants. Although Bartram did not live with the natives, he traveled extensively in their territory, and his final account of them in part 4 stems from observations he made in the spirit of this pre-anthropological anthropology.

His attempts to depict the natives fall into two groups—scenes in which the depiction is influenced by literary models, and scenes that have a natural historical basis in that they include manners-and-customs portraits of the villagers.[73] The other alternative, a narrative in which the people appear as characters, in much the same way that Bartram is a character in *Travels*, never comes to be written.

Bartram's first encounter with a native American in the narrative portion of *Travels* has a decidedly literary cast. Bartram had ridden beyond the frontier white settlement when he crossed the river St. Ille (now called the Satilla) in southeastern Georgia. Traveling south he saw a mounted, armed American Indian cross the path some distance in front of him. Bartram tried to hide behind some trees, but the "intrepid Siminole" saw Bartram and "sat spurs to his horse."[74] Bartram, unarmed, did not run, but "resigned [him]self entirely to the will of the Almighty." Then Bartram offered the man his hand, and after a bad moment when it seemed that this offering would be met with an angry response, Bartram had the satisfaction of having the man shake his hand in friendship. Bartram asked directions to a nearby trading house, and they parted.

Bartram does not offer the reader an account of the exchange of questions and information—What language did they speak? What was said? How were the directions given? Instead Bartram assumes temporarily the powers of an omniscient narrator and tells us what the other man was thinking:

> Possibly the silent language of his soul, during the moment of suspense (for I believe his design was to kill me when he first came up) was after this manner: "White man, thou art my enemy, and thou and thy brethren may have killed mine; yet it may not be so, and even were that the case, thou are now alone, and in my power. Live; the Great Spirit forbids me to touch thy life; go to thy brethren, tell them thou sawest an Indian in the forests, who knew how to be humane and compassionate.[75]

Bartram writes the man's interior monologue rather than reporting his words. The model here is the "noble savage."[76] The language of the man's supposed thoughts is lofty, literate, and Quaker in its use of pronouns.

Sometimes the literary model for the noble savage is quite

specific. Bartram's account of his meeting with the White King of Talahasochte ends with the king's being quoted as saying, "Our whole country is before you, where you may range about at pleasure, gather physic plants and flowers, and every other production."[77] The chief had been reading Milton.[78]

Christianity is not the only source for this mythologizing of the natives. Bartram and his traveling companions came upon women gathering wild strawberries. Bartram labels this a "sylvan scene of primitive innocence," a "gay assembly of hamadryades." He leaves it to the reader to speculate as to the lengths to which the travelers' "passions might have hurried us, thus warmed and exited" by the sight of these young maidens "wantonly chasing their companions, tantalising them, staining their lips and cheeks with the rich fruit."[79] The travelers come upon these "nymphs" in the "Elysian fields."[80] In this scene the literary transformation of the American Indians is complete. These women are not human at all, but the very spirits of the trees surrounding the travelers.

Bartram provides us with instances of the noble savage's opposite, too. Another American Indian appears as the wife of a white trader. She has robbed him of his possessions, and he is taking refuge in brandy. Bartram notes her beauty, her engaging manners, the "perfection in her person," and calls her husband "her beguiled and vanquished lover and unhappy slave." The situation is described in a summary paragraph, and Bartram includes it, he says, "to exhibit an instance of the power of beauty in a savage, and her art and finesse in improving it to her private ends." Bartram engages an American Indian to travel upriver with him but must put the man ashore after he becomes weary of rowing. Bartram complies, he says, because he knows "the impossibility of compelling an Indian against his own inclinations, or even prevailing upon him by reasonable arguments, when labour is in the question."[81] This American Indian is unnamed, the narrative of his abandonment told in the course of Bartram's pursuit of the natural history that is most often the aim of his voyage.

If the literary models furnished Bartram with both noble

and bad natives, the natural historical models turn the American Indian into the Other. Ultimately, Bartram represents native Americans with exactly the same rhetoric he has been using to represent the plants and animals he encounters on his trip.

Bartram's account of his reception in Cuscowilla, a Creek town in Florida that he visited with four men who were acting as traders' representatives to the American Indians, is unusual in *Travels* in that it provides a fairly complete picture of the encounter. The account is narrative, lapsing from the past tense used for the party's greeting and walk to the chief's house into the ethnographic present for the explanation of the pipe-smoking and communal-drinking ceremony, back into the past tense for a description of Cowkeeper, the chief, and once again into the present for the brief account of the "banquet."[82] The description of Cuscowilla is preceded by a natural history essay on the turtle known (then and now) as the gopher, and a paragraph of description of the plant productions of the marshes near Lake Cuscowilla. It is followed by a description of a level, grassy plain and the Alachua savanna. The description of the community is one more natural history essay sandwiched between the turtle and the savanna. Aside from Cowkeeper's greeting ("You are come") and his dubbing of Bartram "Puc Puggy" (the Flower Hunter), the account is undifferentiated from the natural history essays around it.

It is undifferentiated because it is one with them. Indian chiefs are individualized. "Untitled" townspeople in the scene are simply "young men and Maidens" or "women and children." The account of the meal is a list: "The repast is now brought in, consisting of venison, stewed with bear's oil, fresh corn cakes, milk, and homony; and our drink, honey and water, very cool and agreeable."[83] The tense is the ethnographic present. The list is one with the account of the plain: "After partaking of this banquet, we took leave and departed the great savanna. We soon entered a level, grassy plain, interspersed with low, spreading, three-leaved Pine-trees, large patches of low shrubs, consisting of *Prinos glaber,* low *Myrica, Kalmia galuca,*

Andromedas of several species, and many other shrubs, with patches of Palmetto."[84] The American Indians are treated as just one more item of natural history.

After the travelers return from the savanna, they conclude their business with the native Americans, and Bartram has another opportunity to present the reader with an account of this town and its inhabitants. He provides us with a long description of the "habitations," and of the town's situation on the edge of the lake, again presented in the historical present. A description of the "plantation" (common fields), including a Linnaean list of the plants the townspeople grow there, concludes Bartram's second look at Cuscowilla. The description of the town is as static as the description of the nearby savanna. Bartram employs this descriptive method when he describes a little (unnamed) village, and the Seminole towns of Sinica and Cowe.[85] The long description of Cowe eventually gives way to a list of forty-three "towns and villages in the Cherokee nation inhabited at this day."[86]

At Attasse, another American Indian town, Bartram spends a week waiting to travel to Augusta with some traders. He offers us an extensive description of the town itself and of the inhabitants' "vigils" and "vespers." He describes the buildings, and speculates as to the means of construction that must have been used to erect a huge pine pole in a mound of earth, when no such pine grows within twelve miles of the place.[87] This is the most extended description of a single native community that appears in the narrative portion of the *Travels*. In it not one person is individualized, not one distinguished from the rest. And once again, the description becomes a list, as Bartram recites the towns on the Tallapoose, Coosau, Apalachucla, and Flint rivers, along with the languages spoken in each town.[88]

In his treatment of the chief of Whatoga, Bartram mixes the method of natural history with a rare instance of personal narrative. He finds himself at the edge of a town planted with separate plots that adjoin one another in such a way that there are only narrow spaces between them to allow passage for his horse, "when observing an Indian man at the door of his habitation, three or four hundred yards distance from me, beckoning me to come to him, I ventured to ride through their lots, being careful to do no injury to the young plants, the rising hopes of their labour and industry."[89] Bartram's narrative carries us with him to the man's home, where we have described for us the meal the chief served him, the calumet they shared, and the corn ordered for Bartram's horse. The chief walks with Bartram two miles out of town to show him the right course for Cowe, Bartram's next destination. The narrative shows us the actions of a single man but ends in a passage with the familiar ring of natural history in which the individual fades into the representative male:

> This prince is the chief of Whatoga, a man universally beloved, and particularly esteemed by the whites for his pacific and equitable disposition, and revered by all for his exemplary virtues, just, moderate, magnanimous and intrepid.
>
> He was tall and perfectly formed; his countenance cheerful and lofty, and at the same time truly characteristic of the red men, that is, the brow ferocious, and the eye active, piercing or fiery, as an eagle. He appeared to be about sixty years of age, yet upright and muscular, and his limbs active as youth.[90]

For Bartram there are few real choices for him in his representation of the American Indians. His heroic models from literature betray him into moral flights of eloquence. His single attempt at extended narrative ends in an idealized portrait of a great and heroic chief. His portrait of the deceitful wife of a European trader comes closest to a depiction of an individual, perhaps because she had distinguished herself to Bartram in her choice of husbands, or in her lapses in morality.

In the end, he turns to the rhetoric of natural history, offering in part 4 "An Account of the Person, Manners, Customs and Government of the Muscogulges or Creeks, Cherokees, Chactaws, &c., Aboriginies of the Continent of North America."[91] Here he presents a standard manners-and-customs account still consulted for its accurate depiction of American Indian characteristics. In 1789 Bartram wrote a similar account in response to questions put to him by Benjamin Smith Barton; the American Ethnological Society published it in 1853 in its *Transactions*.[92]

To present an adequate account of the southeastern natives, Bartram removed them from the narrative portion of *Travels*. In part 4 his actions as traveler are no longer the framing device that they were for parts 1 through 3; part 3 ends with Bartram's return to his father's house. He cast part 4 in the eternal now of the historical present. The final portion of Bartram's text represents these people as without a history, either as a group or as individuals, while it furnishes readers with thematically organized descriptions of their persons, character, government, dress, feasts, property, agriculture, arts, manufactures, religion, language, and monuments. Bartram visited the southeastern natives as an observer, with all of the detachment that such a role implies.

Why were the native Americans banished from the narrative when the plants and other animals that Bartram encountered were included, however disruptively, in the portion of the book framed and punctuated by Bartram's account of his movement through the Southeast? The same method and rhetoric produced both sets of descriptions. But in the American Indians Bartram faced fellow humans. His method for describing these people was first formulated to describe plants and animals that were clearly not of the same species as the describer. A tree is clearly Other. So is a bird. Transfer the method used to describe the tree or bird to another human being, and he becomes Other, too. The observer is led to see only what the method sanctions. He is led to record only what the rhetoric has devised formulas to represent.

The Southeast, and by extension the America represented in *Travels*, is a place where objects can be named using the Linnaean system, and, conversely, a place where Linnaean names have referents. It is a part of the universal system, a place where natural things are found that are related to the ones in the reader's own country. It is a room in the "glorious apartment of the boundless palace of the sovereign Creator."[93] The Great Chain provides Bartram and his reader with a shared system for arraying the things that the traveler finds and reports.

Bartram himself provides a second, parallel system for arraying the objects. The plants and animals of the Southeast accumulate before the reader's very eyes as their Linnaean names are called: *Magnolia grandiflora, Myrica cerifera, Laurus borbonia, Rhamnus frangula.* The discrete things denoted by the Latin names arrange themselves, through accretion in list after list, into a series of southeastern locales: the seacoast island, the riverbank, the savanna, the forest, the swamp, the inland lake. In a series of epitomes made possible by and conveyed almost solely through Linnaean nomenclature, Bartram painstakingly represents the country through which he traveled for almost four years. The effect is visual and scenic, although Bartram does not construct landscapes. For Bartram, the Southeast is a locale that exists prior to the kind of art that the picturesque grew from—landscapes sought out and arranged according to aesthetic rules. The American locales that Bartram sought and represented were unmediated by such artifices.

In addition to telling the reader what the scenes looked like, Bartram provides his own reactions, but not before he has demonstrated the theory behind those reactions—the sublime as defined by Edmund Burke. For Burke, and so for Bartram, one observer's reactions were those that any observer might have.

Yet the scenes, like many of the plants and animals in *Travels,* were new. Many of them had never been described. Both the method of natural history and the theory of Edmund Burke provided the observer with a shared context within which he could circumscribe the novel objects and scenes he encountered in the Southeast. The objects there are like objects the reader is familiar with, and they have their places on the Great Chain. The observer there is like observers everywhere, moved to awe and terror by the novel scenes that make up the Southeast.

Thus America, simultaneously exotic and familiar, is represented to the sight and emotions of the reader.

The method makes the land appear new; it strips the land of its history. Natural historical description is static and atemporal. As represented by the method, the Southeast is a place where there are no verbs. Thus Bartram's America seems to be waiting for history to happen and for individuals to live their lives there. The remains of Spanish settlements are represented by means of the same rhetoric that Bartram uses to describe the plants or the soil of a given locale.[94] American Indians, also represented by the same rhetoric, are not individuals with a personal history, nor are they a group with a collective history.

Ultimately, the narrator is not even a character in his own narrative. In 1766, Bartram tried and failed to make a living on his own indigo plantation in East Florida.[95] On the subject of the author's year-long residency in the country he described, *Travels* is absolutely silent. Any such personal relationship to the land is unrepresentable by the natural historical method and its attendant rhetoric, which requires contemplation by both observer and reader, but makes no provision for use. Bartram moves through a fully furnished American Southeast with no role to play other than that of observer and recorder. He leaves, in *Travels,* a record but not a memorial.

In his descriptions America is a vast, still garden, planted and denizened with species that have names but no one to use them. It is known, but not truly inhabited.

• CHAPTER THREE •

Jefferson and the Department of Man

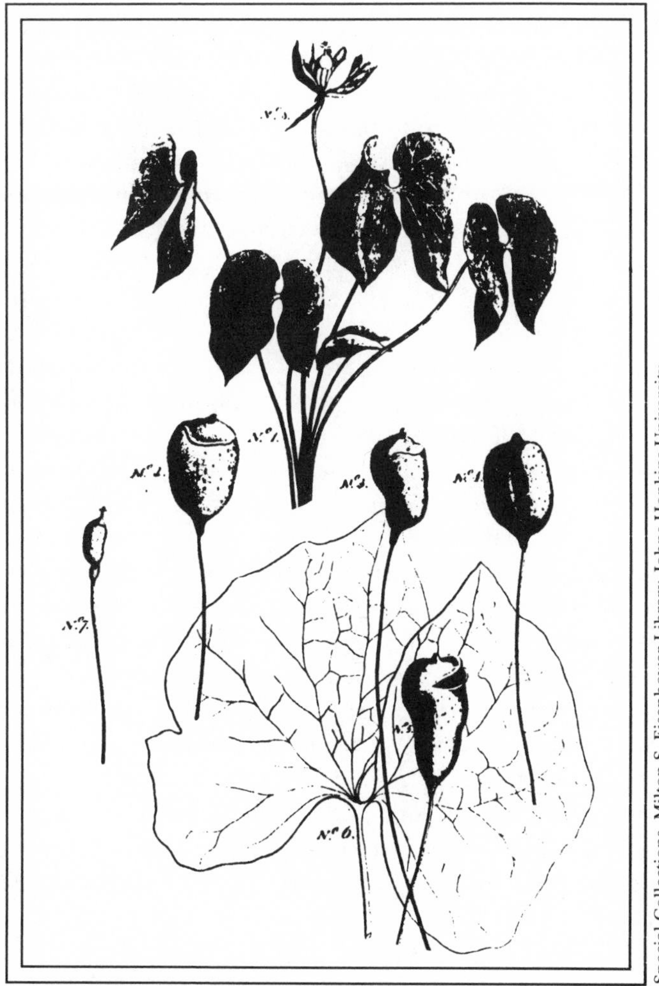

Special Collections, Milton S. Eisenhower Library, Johns Hopkins University.

In 1792, to commemorate Thomas Jefferson's knowledge of natural history, Benjamin Smith Barton named this plant Jeffersonia binata. ***Jefferson*** *planted the native Virginian species in his garden at Monticello.*

In a letter to David Rittenhouse, preeminent colonial mathematician and builder of an orrery much celebrated by his Newtonian contemporaries, Thomas Jefferson scolded him for his employment in the governing of Pennsylvania, calling it "commonplace drudgery . . . a work which may be executed by men of an ordinary stature, such as are always and everywhere to be found."[1] Rittenhouse's real vocation was a higher one—astronomy and mathematics. Jefferson believed that natural history also summoned its practitioners to that higher plane, one that transcended the drudgery of politics, the flux of mere events. *Notes on the State of Virginia* is Jefferson's most public record of the scientific observations he made throughout his life. The book was written in response to a series of questions circulated in 1780 by François Marbois, the secretary of the French legation at Philadelphia.[2] The diplomat sought information about each American state. Jefferson was deemed best able to answer for Virginia. Taken together, his responses define, delineate, number, list, and describe his "country," beginning with its boundaries and ending with its state papers. First circulated in manuscript, Jefferson's replies were corrected and enlarged, then finally printed in 1785.[3] Seen as an instance of the literature of place, *Notes* can be appreciated for its comprehensive representation of Virginia, an accounting that Jefferson updated during the twenty years following its publication.[4] The book's organization, method, and rhetoric are scientific. Natural history's primary activities—observation, description, and careful record keeping—were essential tools of Jefferson's wide-ranging genius, and natural history is the primary intellectual framework on which he built *Notes*.

The most recent historian of ideas to examine Jefferson as a scientist, John C. Greene, dismisses his scientific career as "unimpressive" and emphasizes his intellectual leadership in promoting the scientific explorations of others.[5] In tracing Jefferson's administrative contributions to scientific inquiry, Greene treats natural history in the light of its successors. He presents Jefferson's considerable grounding in natural history not as the mastery of a single science, but as a series of tentative beginnings in those disciplines that would gradually lay claim to natural history's intellectual territory— archaeology, geology, and the like. In his account, Jefferson's contributions are scattered, his intellectual orientation fragmented. Edwin T. Martin, writing almost three decades earlier than Greene, focuses on Jefferson's mechanical and architectural innovations, such as the moldboard for a plow of least resistance, the clock in the main hall at Monticello, as well as the house itself and the original campus of the University of Virginia, all of which Jefferson designed.[6] Martin, in emphasizing Jefferson's mechanical and architectural achievements, highlights his contribution to branches of knowledge that still survive. Both approaches "read" Jefferson's thought in light of what has endured, a process that appears reasonable when the thinker in question has written the most enduring document in American history. However, in adopting this goal of demonstrating Jefferson's modern relevance, these historians read Jefferson's *Notes* from a modern perspective. They overlook the science in the text that does not resonate in the mind of the modern reader, or they dissect that which was, for Jefferson, entire, diffusing the unity of the whole of natural history into pieces of occasional significance separated by long, irrelevant passages.

Given this approach on the part of historians of science, it is not surprising that literary critics have also read selectively, concentrating on a few passages in the work that they could identify as belletristic. They have ignored much of the rest, or, more damaging, generalized from brief, unrepresentative passages. In the 300-page work, just three paragraphs—which anthologists have turned into Jeffersonian set pieces—have received the bulk of critics' attention.[7] Concentration on the statement of the agrarian ideal in Query XIX ("Those who labor in the earth are the chosen people of God, if ever he had a chosen people") and on the descriptions of the Blue Ridge Gap (Query IV) and the Natural Bridge (Query V) constitutes a misrepresentative criticism that leaves most of the book unread.

Commentators on these passages typically return to the idea of "nature," yet they do so without truly crediting the natural historical context from which Jefferson makes his observations. Instead, they dismiss the natural historical basis of *Notes*.[8] Jefferson was perhaps the most accomplished of the writers who strove to represent America, or a part of it, within the covers of a book. Natural history is the key to that enterprise. It contributes an organization and a rhetoric to the *Notes*, and Jefferson as a practitioner of science writes a book that describes Virginia, his country, comprehensively. Yet in his representation of blacks and American Indians, what he calls "the department of man," natural history fails Jefferson.[9] He recognizes the inadequacy of his means of representation but is unable to devise a better one. This failure foreshadows in his text the future misfortunes of these two great casualties of America's founding.

Jefferson was born into a family interested in natural history. Jefferson's maternal grandfather, "Captain" Isham Randolph, was tied into the natural history circle through his interest in botany and his correspondence with Peter Collinson and John Bartram.[10] Jefferson's father, Peter, was a landowner and surveyor who drew an early map of the province on which Jefferson based the map he included in the *Notes*.[11] Details of Jefferson's education are sketchy. His first library and copies of his correspondence were lost in 1770 when Shadwell, which Jefferson had inherited from his father, burned. Jefferson was twenty-seven.[12] Six years later he would write the Declaration of Independence; ten years later, *Notes on the State of Virginia*. Although records of Jefferson's reading during his formative years perished in the fire, he did leave a testimonial to the two teachers who most influenced him. One was William Small, who taught natural philosophy at William and Mary.[13] Unlike the other teachers in the philosophy school, Small's methodology was scientific.[14] After leaving William and Mary, Jefferson read law under the other of his "most influential" teachers,

George Wythe. As governor, Jefferson created at William and Mary professorships in both law, and anatomy and medicine, an early instance of the encouragement of science for which he would later be celebrated, and a reflection of the respect he had for these subjects.[15]

In 1764, when he assumed control over his inheritance, Jefferson owned or managed approximately 5,000 acres of land.[16] His holdings eventually exceeded 10,000 acres, on which lived and worked more than 100 slaves.[17] In addition to raising crops for market, chiefly tobacco and wheat, Jefferson grew food for his family, workers, and slaves. In his garden records he listed thirty different names of cabbage, over forty of beans, and fifty of peas, as well as more than seventy other vegetables and herbs.[18] The number of varieties indicates his interest in garden plants for their own sakes. This practical botanical bent extended to his farmland and pastures as well. During his long absences from his homes, he planned and directed the improvement and planting of his land.[19]

To promote the advancement of American agriculture and to satisfy his own curiosity, Jefferson was himself a member of the natural history circle, carrying on an active botanical correspondence with men such as John Bartram, Jr., the inheritor of his family's garden; Benjamin Smith Barton, author of the first American botanical textbook; Frederick Pursh, describer of plants collected by the Lewis and Clark expedition; André Thoüin, student of Buffon and head gardener at the Jardin des Plantes in Paris; Baron F. W. H. Alexander von Humboldt, pioneering climatologist and widely traveled natural historical observer and collector; and Dr. Benjamin Say, nephew and student of William Bartram and pioneer of the study of entomology in America.[20] In Paris, serving as minister to France, Jefferson became a clearinghouse for seeds and scientific information of all kinds, including copies of a new English translation of Linnaeus's *Systema vegetabilium.* Soon after his return to America, he paid a subscription for Bartram's *Travels.* As secretary of state under Washington he lived across the Schuylkill from Bartram's garden, and was a frequent visitor.

At Monticello Jefferson planted *Jeffersonia diphylla,* which Benjamin Smith Barton had named in his honor.[21] He knew the work of Miller and Clayton, in addition to that of Linnaeus. From 1797 through 1814 he served as the president of the American Philosophical Society, to which he had been elected in 1780, just one year before he drafted *Notes.*[22]

Jefferson's interest in plants united his landholding responsibilities with his scientific bent, but he also studied virtually every other natural production. He made pioneering observations of the Virginia climate. He supported the excavations of mammoth bones by Charles Willson Peale. He himself excavated an American Indian barrow, employing methods that archaeologists use today. Believing that knowledge of American Indian origins would emerge from linguistic comparison, he compiled vocabulary lists of various native languages. He proposed and sponsored the Lewis and Clark expedition, first sending Meriwether Lewis to Philadelphia where Benjamin Smith Barton, Caspar Wistar, and Benjamin Rush could train him in natural history.[23] In 1809 he wrote to E. I. du Pont de Nemours, "Nature intended me for the tranquil pursuits of science."[24]

Learned in virtually every field, Jefferson had ample knowledge of science, particularly natural history. In one letter, he provided a working definition of this now-superseded field of inquiry: "as soon as the structure of any natural production is destroyed by art, it ceases to be a subject of natural history, and enters into the domain ascribed to chemistry, to pharmacy, to anatomy, etc."[25] Jefferson had little use for these more analytic fields. He was not an experimentalist. He was an observer, and the *Notes* is a work of informed observation and meticulous record keeping of natural historical phenomena.

In Jefferson's Baconian division of knowledge, natural history relied on the mental faculty of memory. Moral philosophy, which included ethics and the law, was a function of reason.[26] Memory and reason are the realms of knowledge through which

Jefferson ranged in the *Notes*.[27] Robert A. Ferguson has trac the influence of the law and legal reasoning in Jefferson's text.[28] In making his case for the importance of the law, however, Ferguson claims that Jefferson rejects the organizing power of natural history because the New World was "too remote" to yield itself to natural historical description.[29] If Jefferson's *Notes* marks the beginning of a belletristic American literary tradition dominated by lawyers and shaped by the kind of training that lawyers received, it simultaneously occupies an important place in the tradition of texts that constitute the literature of place. When Ferguson asserts that the law, a branch of moral philosophy, organizes the *Notes* and provides it with a rhetoric, he slights the role played by natural history in the work, the importance of memory as well as reason in Jefferson's book. For Jefferson, neither memory (and its product, information) nor reason could do without the other.

Ferguson asserts that Jefferson could not use natural history as an ordering principle in *Notes* because Virginia was but a "half-known world," which Jefferson himself often admits.[30]

> Each admission of an incomplete answer . . . contains Jefferson's realization that the bulk of the New World remains too remote to allow the use of natural philosophy as the primary structural principle for *Notes*. . . . Neither interest in the natural world nor general philosophical acceptance of nature's blueprint for man can compensate the author of *Notes* for his own lack of information.[31]

Ferguson has misunderstood natural history, assuming that in the absence of complete information Jefferson simply could not employ the natural historical method. He could and did. A responsible practitioner of natural history admitted ignorance rather than supplying erroneous information. Jefferson recorded such admissions at appropriate places in *Notes*. Despite these claims of ignorance—the modest statements of an avowed skeptic—Jefferson had a great deal of natural historical information. The longest section of the *Notes*, the answer to

Query VI, is more than twice as long as any of the other twenty-two answers, and this section is on the classic natural historical topics: productions mineral, vegetable and animal. Fully half of the book is about natural history. Jefferson's Virginia may have been only half known, but he told the half that he knew.

That telling employs the methods and rhetoric not only of the law but also of natural history. Ferguson admits this dual governance even in the process of arguing for natural history's lesser importance: "*Excluding section six with its extensive lists of indigenous flora and fauna,* Jefferson gives most space in *Notes* to sections thirteen and fourteen—the central rationale of an American polity based upon charters, constitutions, and the laws of the state. These sections . . . form the heart of the book."[32] Ferguson locates the link between these later sections and the natural history of the first seven sections of the book in Query VII, "Climate":

> Jefferson's seventh section involves a discussion of Virginia's climate based on painstaking tabulations of rainfall, temperature, and wind velocity recorded twice a day across a five-year period. And yet the heading for the section . . . calls far more grandly for 'a notice of all what can increase the progress of human knowledge.' Narrow meteorological expertise seems an inappropriate response to the question. . . . [But] Montesquieu, . . . accepting the general notions that good government should be conformable to nature and that positive law should build from natural law, emphasized the importance of geo- physical elements in describing national identity.[33]

For the eighteenth century, climate helps establish national identity.[34] Jefferson's discussion of it links his account of Virginia's natural history to the later sections on government and law. It also serves another function: Climate has as its essential constituent "vital air," the "immediate cause or primum mobile of life."[35] In this note to Query VII Jefferson traces the effects of heat and cold on the human body, and he includes a clinical explanation of the workings of what would come to be called the cardiovascular system, ending with "the body becom[ing] an inanimated lump of matter" when "vital air" is withheld from the lungs. For the eighteenth century, climate also contributes to the very presence or absence of the life that enables people to forge such a national identity. With Jefferson, as with Bartram, nature was teleological. The force of causality held sway from the primum mobile of the human body to the characteristics of the body politic. Natural history was one of the foundations upon which all other histories were built.

This foundation is reflected in the organization of the *Notes*. Jefferson reordered the list of questions that Marbois submitted to him, changing their sequence and moving certain topics from one question to another. For Ferguson, Jefferson's reordering reflects the influence of eighteenth-century writers on the law: Grotius, Pufendorf, Burlamaqui, Montesquieu, and Blackstone. It embodies "an elementary structural separation between natural phenomenon and social event."[36] This separation between natural history and social event is not as easy to mark as Ferguson believes. The first six sections—"Boundaries"; "Rivers"; "Sea-Ports"; "Mountains"; "Cascades"; and "Productions Mineral, Vegetable, and Animal"—define a territory ("Boundaries") and describe its geographical features, its flora, and fauna. The next six sections—"Climate," "Population," "Military Force," "Marine Force," "Aboriginies," and "Counties and Towns"—define the condition of the primum mobile of life ("Climate") and number the humans who breathe America's air. Only in the last eleven sections—"Constitution"; "Laws"; "Colleges"; Buildings and Roads"; "Proceedings as to Tories"; "Religion"; "Manners"; "Subjects of Commerce"; "Weights, Measures, and Money"; "Public Revenue and Expences"; and "Histories, Memorials, and State-Papers"—is social event introduced. Each of these three divisions is governed by an introductory section that forms the basis for the sections that follow—"Boundaries" circumscribes natural history, "Climate" circumscribes the department of man, and "Constitution" circumscribes social event. If one views human inhabitants as necessarily a part of social event, then only six out of twenty-three chapters are about natural history. If, however, one views human inhabitants as the last animal link on the Great Chain, then the number of sections about natural history increases to twelve out of twenty-three. Natural history forms the setting for social event. Social event is the natural behavior of that species of animal, *Homo sapiens,* that occupies the top of the Great Chain of Being. Jefferson's primary concern in the chapters between "Climate" and "Constitution" is to record the number of people of various kinds, both within and without Virginia's boundaries, a procedure more closely allied to natural history than to the representation of social event. Jefferson's organization owes much to both natural history and natural law.

Natural history contributes as much to the method and rhetoric of the *Notes* as does the law, and in similar ways. Ferguson cites the method of "legal examination," which Jefferson learned as he read law, to account for the rhetoric of his book: "The very structure of *Notes*—an order through accumulation, leaving room for later elaboration—parallels that of the common law."[37] "Order through accumulation" could as easily describe the structure of Linnaeus's works. It is the *modus operandi* of natural history as well as of the common law. The causes of innovation in both intellectual spheres can be traced to novelty. New laws are written, or old ones are given new interpretations, when novel human circumstances arise for which existing laws and interpretations do not adequately account. New natural historical descriptions and their names are added to the body of natural historical knowledge when novel natural productions are discovered that cannot be adequately characterized by existing descriptions.

The method of natural history required by Linnaean nomenclature, a method Jefferson put to work in the *Notes,* is similar to the analytic method used by lawyers then and now. In entries in his *Commonplace Book,* Jefferson outlines the difference between trespassing and the "breaking" required for a theft to be counted as a burglary:

> Every entry into a house by a trespasser is a breaking in law: as if the door of a mansion house be open, and the theif enter with a purpose to steal, this is a breaking in law. But in cases of burglary, an actual breaking is necessary, if the window of a house be open, and a theif with a hook draw out some of the goods of the owner, it is not burglary: but otherwise if he break the window, when a theif has broke the house, an entry of any part of his body, or of a gun, or hook, into the hole which he hath made with an intent to steal or kill, is such an entry as will constitute burglary.[38]

The lawyer begins with a string of events, a narrative, presented to him by a client. For example: On a rainy night, under the cover of fog, the client walked up to the house of another man, uninvited, and knocked on the door. When no one answered, he forced the lock, entered the house, and took the other man's silver. In deciding what this client can be charged with, the lawyer eliminates the elements of the narrative that do not help him apply or reject a given term of art, in this example, "burglary." The rain and fog are incidental and can be safely ignored. Walking on another person's land uninvited is essential; it constitutes trespass. Forcing the lock is the "actual breaking"; taking the silver is the theft. All of the elements of burglary are present, and this client can be charged with that crime. During the five years that Jefferson read to become a lawyer, he learned to analyze narratives in just this way.

Compare this method to the one embodied in Jefferson's musings on the platypus in a letter to Dr. John Manners:

> Although without mammae, naturalists are obliged to place it in the class of mammiferae; and Blumenbach, particularly, arranges it in his order of Palmipeds and toothless genus, with the walrus and manatie. In Linnaeus' system, it might be inserted as a new genus between the anteater and the manis, in the order of Bruta. It seems, in truth, to have stronger relations with that class than any other in the construction of the heart, its red and warm blood, hairy integuments, in being quadruped and viviparous, and may we not say, in its *tout ensemble,* which Buffon makes his soul principle

> rangement? The mandible, as you observe, would draw it ırd the birds, were not this characteristic overbalanced by the ghtier ones before mentioned. That of the Cloaca is equivocal, cause although a character of birds, yet some mammalia, as ıe beaver and the sloth, have the rectum and urinary passage terminating at a common opening. Its ribs also, by their number and structure, are nearer those of the bird than of the mammalia. It is possible that further opportunities of examination may discover the mammae.[39]

In natural history as in law, Jefferson's method involves discovering what is essential, ignoring what is not, and applying the correct label. The natural historian begins with a specimen, in this case the platypus. He then considers the specimen's affinities with animals that have already been placed in the Great Chain—its warm blood and the construction of its heart—which argue for a placement in a given order (Bruta) between two other animals that already form links on the Chain. He also considers another element, the mandible, which seems to argue for placement nearer the birds but finally seems inessential, however remarkable it might be. At the heart of both methods, natural historical and legal, is classification. Certain elements argue for the appropriateness of a given label, others need to be disregarded. Lawyers classify actions; natural historians classify natural productions. Both reason their way to the essential, eliminating the incidental. Jefferson did not need to turn away from the method of natural history to embrace the method of the law. In essence, they were the same.

Much more than anyone has seen, natural history provides a key for reading *Notes,* a clue to the kind of reader for whom Jefferson was writing, and the kind of rhetoric he was using. Charles Thomson, secretary of the Continental Congress and author of the commentaries to the *Notes* that form appendix 1, called the book "a most excellent Natural History not merely of Virginia but of North America and possibly equal if not superior to that of any Country yet published."[40]

In Query VI, "Productions Mineral, Vegetable, and Animal," Jefferson demonstrates his familiarity with the Linnaean method:

> Paccan, or Illinois nut. Not described by Linnaeus, Millar, or Clayton. Were I to venture to describe this, speaking of the fruit from memory, and of the leaf from plants of two years growth, I should specify it as the *Juglans alba, foliolis lanceolatis, acuminatis, serratis, tomentosis, fructu minore, ovato, compresso, vix insculpto, dulci, putamine, tenerrimo.* It grows on the Illinois, Wabash, Ohio, and Missisipi. It is spoken of by Don Ulloa under the name of Pacanos, in his Noticias Americanas. Entret. 6.[41]

This is a textbook Linnaean description. Jefferson consulted Linnaeus, Miller's *Gardener's Dictionary,* and Clayton's (Gronovius's) *Flora Virginica.* Not finding a published systematic description, he supplied his own.

As we saw in Bartram's *Travels,* a characteristic form emerging from the method of natural history is the list. This is a form that Jefferson found most congenial. *Notes* contains more than twenty lists and tables. Two entire sections are themselves lists—Query II, "Rivers," and Query XXIII, "Histories, Memorials, and State-Papers." Jefferson lists minerals; vegetables; the quadrupeds of America and of Europe, in addition to a comparative list of both American and European quadrupeds; birds; winds; population; potential increase in population with and without new immigrants; taxable articles; militia; native Americans northward and westward of the United States as well as those within the United States; American Indian tribes in Virginia; towns; fighting men; crimes and punishments; exports; rates of exchange; and expenses of government.[42] He refers to one list as a "comprehensive view of the Quadrupeds of Europe and America, presenting them to the eye."[43] His goal in this table is to demonstrate the lack of a consistent difference in size between New World and European animals of the same or similar species. He assumes that the reader will be able to visualize the comparison. Foucault identifies seeing as the essential act in applying the natural historical.[44]

In half of his lists and tables, Jefferson records the results of weighing, counting, or otherwise measuring the temperature, weight, or number of some quality, meteorological phenomenon, or creature. At bottom, the Linnaean method is a counting method, the numbers of stamens and styles determining the place that a given plant will occupy in the table of the world's vegetable productions. For the eighteenth century, the importance of quantification is not limited to the revolution in plant taxonomy. Garry Wills attributes this quantifying urge to the Newtonian, mechanical spirit of the age that permeated all disciplines.[45] Jefferson was a confirmed measurer and calculator. His plantations, with their large number of dependents and vast acreage, required a calculating director to plan for the feeding and employment of the people and animals and for the planting of the land. This calculating bent had theoretical, as well as practical, applications. The most famous arithmetical proof in *Notes* is Jefferson's computation of the number of geniuses per capita that it is reasonable to expect from France or Great Britain, given America's ratio of geniuses (three—Washington, Franklin, and Rittenhouse) to her population (three million).[46] For Jefferson, quantification was a way to knowledge. For Jefferson's age, it was a distillation of reality, a means of representation essentially like the Linnaean: a universal language (arithmetic) applied to the unique situation of the New World.

In the *Notes,* Jefferson betrays an awareness of the overarching scheme of which the lists are representations. In constructing his table of American quadrupeds, he begins with the mammoth, knowing that the inclusion of an animal for whom the only evidence is a pile of bones will attract questions. He raises the objection, then offers this answer to it:

> Such is the œconomy of nature, that no instance can be produced of her having permitted any one race of her animals to become extinct; of her having formed any link in her great work so weak

> as to be broke. To add to this, the traditionary testimony of the Indians that this animal still exists in the northern and western parts of America, would be adding the light of a taper to that of the meridian sun.[47]

In this instance, Jefferson's belief in the Great Chain of Being, a belief illumined with the light of the meridian sun, led him to assert the mammoth's contemporaneous existence. In the instance of the lists themselves, the presence of the Chain in the worldview of the writer allowed him to tell what he did know about America in a form that would be understood and automatically put into a universal context by readers who also shared that worldview. Jefferson's account did not have to be exhaustive because the system that encompassed it, the Great Chain, was complete, if unfinished. The lists and their implied context let Jefferson's readers know that in America, as in the rest of the world, this system was in place and operating.

In the *Notes,* Jefferson is willing to reason his way to new conclusions based on the evidence. Postulating the existence of the mammoth in the western reaches of the continent is just such a conclusion, founded on information, the mammoth bones themselves, and on the system, the permanence of the links in the Chain. In fact, one characteristic form of discourse in the *Notes* is the argument or proof, the most famous of which is Jefferson's refutation of Buffon. Phenomena, for Jefferson, are not isolated. An American otter might weigh 12 pounds, and a European otter, 8.9 pounds.[48] These two observations put side by side suggest something about Buffon's opinions of American degeneracy. In following such data to the conclusions they seem to suggest, Jefferson is still in the natural history mainstream. In replying to Buffon, he was answering one of the greatest natural historians in the world.

In the *Notes,* Jefferson is a skeptic. For him, "ignorance is preferable to error," and he understands that perception is sometimes distorted. The phenomenon of "looming" is one that he has observed from the height of Monticello. A cone-shaped mountain forty miles away "sometimes subsides almost totally into the horizon; sometimes rises more acute and more elevated, sometimes it is hemispherical, and sometimes its sides are perpendicular, its top flat, and as broad as its base." He also understands that instruments may sometimes be flawed. He explains a difference in readings that ought to agree by "a difference of instruments."[49]

Yet this skepticism should not be confused with positivism. As we have seen in the case of the mammoth, Jefferson is more than willing to reason his way to conclusions for which no eyewitness evidence exists. Yet sometimes this procedure results in a conclusion that we can now recognize as a fact. Jefferson notes that pumice is reported floating on the Mississippi. This has "induced a conjecture that there was a volcano on some of its waters." However, only the Missouri River has still not been traced to its source, and there are no volcanoes on the known portions of that river. "[N]o volcano having ever yet been known at such a distance from the sea, we must rather suppose that this floating substance has been erroneously deemed Pumice."[50] Jefferson is a skeptic within the confines of eighteenth-century natural historical practice.

The natural historical method imposes upon Jefferson's *Notes* the same timeless, static framework we have observed in Bartram's book. Jefferson acknowledges that Virginia's climate has changed "within the memory even of the middle-aged," and he realizes that "a separation [of a single language] into dialects may be the work of a few ages only."[51] Yet his is still an America where the mammoth roams, and where the undiscovered landscape lies in the same state it has always assumed, unchanged by the American Indians ("I know of no such thing existing as an Indian monument"), its natural history waiting to be ordered into lists. In Query XXIII, "Histories, Memorials, and State-Papers," he reduces civil history, the accounts of civilizations and of countries, to a list. Even the events that have left "monuments" are flattened into atemporal catalogues, subservient to the method and rhetoric of natural history.

In the same letter to Dr. Manners in which Jefferson de-

scribed the platypus, he explained the origin of the Linnaea
system:

> Nature has, in truth, produced units only through all her works. Classes, orders, genera, species, are not of her work. Her creation is of individuals. No two animals are exactly alike; no two plants, nor even two leaves or blades of grass; no two crystallizations. And if we may venture from what is within the cognizance of such organs as ours, to conclude on [sic] that beyond their powers, we must believe that no two particles of matter are of exact resemblance. This infinitude of units or individuals being far beyond the capacity of our memory, we are obliged, in aid of that, to distribute them into masses, throwing into each of these all the individuals which have a certain degree of resemblance; to subdivide these again into smaller groups, according to certain points of dissimilitude observable in them, and so on until we have formed what we call a system of classes, orders, genera and species. In doing this, we fix arbitrarily on such characteristic resemblances and differences as seem to us most prominent and invariable in the several subjects, and most likely to take a strong hold in our memories. Thus Ray formed one classification on such lines of division as struck him most favorably; Klein adopted another; Brisson a third, and other naturalists other designations, till Linnaeus appeared. Fortunately for science, he conceived in the three kingdoms of nature, modes of classification which obtained the approbation of the learned of all nations. His system was accordingly adopted by all, and united all in a general language.[52]

Jefferson seems to realize that the system had been imposed arbitrarily. Yet as he applies it to other races, he assumes that it embodies both design and teleology.

Representing the "department of man," both blacks and American Indians, offers special problems to Jefferson. His "condition" (as he might call it) as a slave owner and his employment of the method of natural history made it inevitable that he would regard blacks as Other. In a letter to another plantation owner he refers to blacks as animals: "[Potatoes and clover] feed every animal on the farm except my negroes."[53]

His familiarity with recent evidence that had provided a link between man and beast, thus making the idea of a Great Chain even more urgent, might also have made this Othering all the more urgent.

Jefferson advocated extending natural historical methods to the human race, or at least to those human races sufficiently Other to warrant such scientific treatment:

> To justify a general conclusion, requires many observations, even where the subject may be submitted to the Anatomical knife, to Optical glasses, to analysis by fire, or by solvents. How much more then where it is a faculty, not a substance, we are examining; where it eludes the research of all the senses; where the conditions of its existence are various and variously combined; where the effects of those which are present or absent bid defiance to calculation; let me add too, as a circumstance of great tenderness, where our conclusion would degrade a whole race of men from the rank in the scale of beings which their Creator may perhaps have given them. To our reproach it must be said, that though for a century and a half we have had under our eyes the races of black and of red men, they have never yet been viewed by us as subjects of natural history.[54]

In admitting that blacks have never been the subjects of scientific scrutiny, Jefferson still maintains that natural history is the preferred means for learning about other races. He knows that the tentative conclusion he offers is unjustified because the "many observations" that such a conclusion requires have not been made. He offers it anyway:

> I advance it therefore as a suspicion only, that the blacks, whether originally a distinct race, or made distinct by time and circumstances, are inferior to the whites in the endowments both of body and mind. It is not against experience to suppose, that different species of the same genus, or varieties of the same species, may possess different qualifications. Will not a lover of natural history then, one who views the gradations in all the races of animals with the eye of philosophy, excuse an effort to keep those in the department of man as distinct as nature has formed them?[55]

One critic has called Jefferson's long passage on the inferiority of blacks an "uneasy excursion in the guise of a 'lover of natural history' through the racial fantasies of the white imagination."[56] Jefferson's excursion might betray the racial fantasies of whites, but for him natural history is no "guise." He does cite evidence, albeit not evidence he has gathered himself. And he reasons from analogy, a standard eighteenth-century natural historical practice. The conclusions he reaches we now label as racist, but the kind of argument he uses has persisted, and is with us still.[57] We would now call his procedure bad science. At the time Jefferson wrote, it was perfectly respectable science.

The discourse on black inferiority uses the method of natural history to demonstrate that blacks are a different race, one that is inferior. They are a different color; "whether it proceeds from the colour of the blood, the colour of the bile, or from that of some other secretion," the difference is "fixed in nature. . . . [T]he fine mixtures of red and white" of the white race are more beautiful, as even the blacks admit. Their secretions differ, as, perhaps, do their "pulmonary apparatuses"; they require less sleep, they are more ardent, they feel affliction less strongly; their reason is inferior, their imaginations "dull, tasteless, and anomalous." They have not bettered themselves in America despite their exposure to white society. A comparison of their situation with that of Roman slaves, many of whom were brilliant, demonstrates that "it is not their condition . . . but nature, which has produced the distinction."[58]

In this litany of facts demonstrating black inferiority, Jefferson mentions "the preference of the Oran-ootan for the Black women over those of his own species."[59] This aside strongly suggests that Jefferson had read Edward Tyson's *Orang-outang sive Homo Sylvestris or the Anatomy of a Pygmie* (London, 1699), a work that made widespread the idea that man and ape might be related.[60] We know that in 1788, six years after Jefferson finished drafting the *Notes,* he ordered "Tyson's Orang-outang, or anatomy of a pigmy" from Thomas Payne, a London bookseller.[61] Tyson is remembered as the father of primatology. One less-than-pristine "fact" included in his work is the supposed sexual preference of a male orangutan for a black woman. Tyson included this information in the first work that pointed out the physical similarities between the apes (represented by the "man of the forest") and man, similarities that are now taken for granted. One commentator on Linnaeus's career notes that

> in the early eighteenth century the comparison of ape and man was fully established, indeed almost inescapable. . . . The idea of a hierarchy and a continuity in nature, of a scale and a chain, gained in popularity, particularly after the higher apes came on the scene. . . . Linnaeus was convinced that neither physical criteria nor even other characteristics permit a boundary to be drawn between man and ape.[62]

For the eighteenth century, the lack of a boundary between *Homo sapiens europaeus* and the apes was an unsettling idea. The discovery of one or more species between the apes and "higher" man could "buffer" *Homo sapiens europaeus* from an unpleasant propinquity to beasts.

In a passage reminiscent of the Abbé Raynal accusing America of a lack of geniuses, Jefferson enumerates blacks' supposed intellectual shortcomings. This reasoning forms the apotheosis of the Othering in which Jefferson has been engaging. Blacks lack every intellectual attainment, including those that distinguish scientists: Ignatius Sancho is guilty of "substituting sentiment for demonstration," which Jefferson adds was also the fault of Buffon.[63] The only solution is emancipation, training, and relocation of blacks in their own homelands. Natural history sanctioned this solution.

Jefferson's effort to control his discussion of blacks ultimately fails, a failure that he recognizes and admits to: "It is impossible to be temperate and to pursue this subject through the various considerations of policy, of morals, of history natural and civil." Blacks eluded classification at every turn: American policy toward them contradicted the "firm basis" on which all other policy was based, that the liberties of a nation are the gift of

God to all people.[64] American morals were corrupted by them. Jefferson's admission of this defeat is placed in the section headed "Manners": "There must doubtless be an unhappy influence on the manners of our people produced by the existence of slavery among us."[65] The law, a part of morals in Jefferson's division of knowledge, also fails.[66] "[L]aws, to be just, must give a reprocation of right," and no such exchange of rights exists between blacks and whites.[67] Natural history cannot help. The evidence is incomplete: blacks have never been made true subjects for natural historical consideration. Civil history also fails. Jefferson laments the difficulty colonial assemblies met with when they tried to tax or halt the importation of slaves into America.[68] In a note he traces this attempt as far back as the "first settlement of Europeans in America," the 1493 Spanish colony in St. Domingo.[69]

Blacks and the institution visited upon them and their masters defy every means of intellectual control at Jefferson's disposal. The only other order available is supernatural: "Indeed I tremble for my country when I reflect that God is just: that his justice cannot sleep for ever: that considering numbers, nature and natural means only, a revolution of the wheel of fortune, an exchange of situation, is among possible events: that it may become probable by supernatural interference!"[70] Natural history and Jefferson's "condition" made the Othering of blacks automatic. But no intellectual means at his command could make their lot seem just. Nature itself, simply through the natural increase in the number of blacks, might grant them numerical superiority over whites, and thus turn the wheel of fortune. If nature alone does not exchange the situation of the two races, Jefferson fears, the Almighty will see that it happens. Natural history, nature itself, and justice are all at odds in Jefferson's formulation, and he has no means to reconcile them.

The other separate race dwelling in America, occupying at the same time its niche in "the department of man," was the American Indian. Jefferson has much greater success in maintaining order in the portions of the text dealing with the original Americans. However, the rhetoric of natural history ultimately proves his undoing as the need to retain order and control forces Jefferson to suppress the one narrative of an individual American Indian that had offered the reader a glimpse of the Other as a human being.

Native Americans were still very much a feature of the countryside near Shadwell, where Jefferson grew up.[71] At William and Mary, Jefferson would have known of the Brafferton, one of the college's professorships and its associated school, which was created to Christianize America's aborigines.[72] When he drafted a bill to reform education in Virginia, he proposed that the school be abolished to be replaced by a missionary-scholar who would minister to the native Americans and study them at the same time.[73] Jefferson's relationship with blacks, as we have seen, was vexed. Ownership of people lay outside the explanatory power of Jefferson's theories of law and natural history. But he inhabited the same land as the American Indians, and it seemed vast enough to accommodate both cultures. For him, natural history had a better chance of explaining America's original inhabitants.

Indeed, Query XI, "Aborigines," is devoted to providing "a description of the Indians established in that state."[74] This section begins with their range and progresses through their manners, history, and monuments. Interspersed are tables that occupy more than a third of the chapter, naming, locating, and counting the natives who lived in Virginia and those who lived "Northward and Westward of the United States."[75] The section is a model of natural historical rhetoric. In its organization, it recapitulates in miniature the entire book. Taken as a whole, *Notes* presents a picture of the country occupied by Virginians, beginning with its boundaries and rivers, progressing (roughly) through its laws, manners, and memorials. The tables of native American tribes, like the tables of American quadrupeds, array America's aborigines such that they appear known and understood in the greater scheme of the world. Jefferson might be unclear as to their precise location on the Chain, but he is undoubtedly clear as to how one should study them. Observe them. Listen to them and record their vocabularies, chiefly so that one might know their origins.[76] Count them. Make lists and tables.

In a famous passage from this chapter, Jefferson recounts his exhumation of an American Indian barrow. His procedure is an early example of systematic archaeology. His findings are described in a long passage that follows the assertion that he knows of "no such thing as an Indian monument" (by which he seems to mean artifacts) and no American Indian "labor on the large scale . . . as respectable as . . . a common ditch."[77] Yet Jefferson is concerned with native origins and speculates as to the antiquity of their inhabitation of America through calculations based on the number of dialects these people spoke.[78]

Jefferson subscribed to the idea that in our language, we carry around with us our origins. He espoused the study of Anglo-Saxon for all university students and as rector of the University of Virginia was the motivating force behind the establishment in 1825 of a course in that language.[79] Believing that "the pure Anglo-Saxon constitutes at this day the basis of our language," Jefferson urged its study to promote "a perfect understanding of the English language."[80] Anglo-Saxon existed in written texts; the American Indian languages, however, existed only in the memories of their speakers. To compile a written record of aboriginal language, Jefferson devised "blank vocabularies" and handed them out to travelers who would quiz the natives and supply American Indian equivalents to English words.[81] His goal was the comparison of the various American Indian languages with one another, and with the Asiatic languages, which had been compiled by Empress Catherine of Russia.[82] If a relationship between native American and Asiatic languages could be demonstrated, the puzzle of the origin of the American Indian might thus be solved.

Jefferson's philological practice differs in the case of Anglo-Saxon and the American Indian languages. Knowledge of Anglo-Saxon is of value to speakers of that language's descendent—to speakers of English. Knowledge of aboriginal languages is sought not for the American Indians' benefit, but to supply the answer to questions about their origins asked by Europeans. Philology, in its application to the American Indian, is allied with natural history. It supplies a description of an aspect of aboriginal peoples that puzzles the investigator, but the study of American Indian languages is never conceived of as a way to provide a history for these people. In Jefferson's "department of man" the idea of origin is permitted, and the immediate description of numbers, location, and extent is an important part of the investigation, but the middle ground between present being and ultimate origin—the ground of history, both individual and societal—is not considered, at least in this section devoted entirely to the American Indians.

It is, however, broached in the section devoted to natural history—Query VI, "Productions Mineral, Vegetable, and Animal." To refute the claim of human degeneracy in America, Jefferson addresses Buffon's assertions about the physical nature of America's inhabitants, with the overall aim of demonstrating the American Indian's ardor. He introduces a section on the their mental qualities by noting that allowances must be made for "circumstances of their situation which call for a display of particular talents only. This done, we shall probably find that they are formed in mind as well as in body, on the same module with the *Homo sapiens Europaeus.*"[83]

To illustrate these qualities of mind, Jefferson relates the story of the Mingo chief Logan. The chief, a resident of the Ohio river frontier, had long been a friend of the whites. But in 1774, when whites murdered his family, he decided to seek revenge. Logan "signalized himself" in the war that ensued when other incidents of white violence against American Indians increased.[84] After the end of Lord Dunmore's War (as the fighting was called), Logan refused to attend the meeting to ratify a peace treaty, but he sent, via a white courier who took down his words verbatim, a message to Governor Dunmore of Virginia that became an immediate classic of native oratory. Newspapers carried it and schoolchildren learned it by heart.[85] This speech Jefferson quotes in full, and he compares it favorably to the orations of Demosthenes and Cicero.[86]

In it Logan explains his actions in the recent war and laments the death of his family. It ends, "Who is there to mourn for Logan?—Not one."[87]

The speech is a set piece, but it and the narrative of the events surrounding it are nonetheless an instance of American Indian history, both individual and national. The speech and the story provide a view of an individual's actions, and because that individual is a leader ("For my country, I rejoice at the beams of peace"), the reader learns a bit of history, a portion of the interactions between whites and American Indians on the frontier. The reader also hears a native American speak, largely because the messenger Logan sent to Dunmore's peace talks took with him a written record of the chief's words. This record was reprinted in the newspapers almost immediately, then copied into account books (like Jefferson's), into diaries, and commonplace books. Thus narrative finds its way into *Notes*. The middle ground between ultimate origins and present being finds representation.

But at the end of the book, in appendix 4, "Relative to the Murder of Logan's Family," Jefferson includes sixteen more tellings of the narrative. Friends of Colonel Cresap, the white man whom Jefferson had held responsible for the murders, had attacked Jefferson's version of the events on the Ohio. The appendix presents the evidence Jefferson gathered to discover the truth or falsity of the narrative he presented in the body of *Notes*. He omits the speech itself but includes fourteen accounts from witnesses of the events surrounding the murder of Logan's family. Then he rewrites the account from *Notes* to better accommodate this new evidence. A final eyewitness account casts Cresap's involvement into doubt. (In fact, it is unlikely that Cresap was responsible.)[88] Jefferson leaves unanswered this challenge to his version of the events. The appendix ends on this inconclusive note.

The long list of retellings betrays narrative by repeating it, making its movement from action to action a single item that itself is repeated. What was once a temporal progression becomes a static value. Logan himself, along with his country, is lost in this narrative of narratives. At work here is the rhetoric of natural history. Logan's tale becomes a sort of specimen, to be compared to other specimens of the same phenomenon so that a type specimen might be extracted therefrom. History, both of an individual and of a "country," is flattened under the weight of this meta-narrative. Temporality, the element that the Linnaean worldview so clearly lacked, is hounded out of the text. The American Indian who had a history and an individual being is gone, replaced by the tabular, atemporal form.

The country represented in *Notes on the State of Virginia* is the static, tableau-like world of natural history and the equally abstracted, static one of the law. If *Notes* was indeed Jefferson's "opportunity for the intellectual discovery of his own country," a "vehicle for the interpretation of his country to himself," what Jefferson discovered when he had ridden the vehicle to its destination was essentially what William Bartram found: a country bounded and denizened, awaiting history.[89]

Jefferson's representation, unlike Bartram's, is numerical. When he referred to the *Notes* as the "measure of a shadow, never stationary, but lengthening as the sun advances, and to be taken anew from hour to hour," he quite self-consciously described the book as a "sketch" of the country at a given "epoch" that provided an "element for calculating the course and motion of this member."[90] "Shadow" implies evanescence. But "measure" fixes the fleeting, provides a stationary quantity in the flux of events. Put together a series of such measurements taken over time, and the country's course can be plotted. For Jefferson, Virginia was a place known in the sense of numbered. For him the wilderness was precise.

Jefferson wrote *Notes* in retreat from the governorship, while Virginia itself was overrun by British troops. His experience as the wartime governor of a state unprepared for military action and the subsequent censure of his executive decisions mounted by his enemies in the Virginia legislature had shaken his dedication to public office.[91] The history being made at the very moment that Jefferson was writing *Notes* was disquieting, and quite unlike the peaceful agrarian society that Jefferson offered as an ideal in his book.

But disquiet enters the static array of the natural historical objects when Jefferson examines the department of man. Blacks elude both natural historical and legal classification. The American Indians, in one section of the book, are mere natural historical objects, their history unspoken, supposed not to exist. In the narrative of Logan, they briefly step forward from the tableau to move and speak, but an appendix reduces that narrative to static fact once again.

On this occasion in the *Notes,* people act rather than simply exist. Their actions—Logan's and Cresap's—accumulate into a story, and that story becomes a narrative in Jefferson's book. As Jefferson's critics were quick to point out, his narrative of these events includes an implication of willful wrongdoing. Cresap and his men "murdered" the American Indians, providing an "unworthy return" of Logan's goodwill toward the whites.[92] This implication caused Jefferson political grief when his enemies attacked him for slandering the name of an innocent white man. Narrative, which is to say, history, incorporates intention and motive, which are, in turn, the essence of politics. History and politics are present in *Notes* only in their attenuated forms—as legal depositions, as verbatim copies of the Articles establishing the colonial governance of Virginia, as a list of problems with the American Constitution. A consequence of its natural historical framework, Jefferson's book obscures the representation of white history, omits the history of blacks, and flattens the telling of a single story about one native American.

• CHAPTER FOUR •

Crèvecoeur's *"Curious observations of the naturalist"*

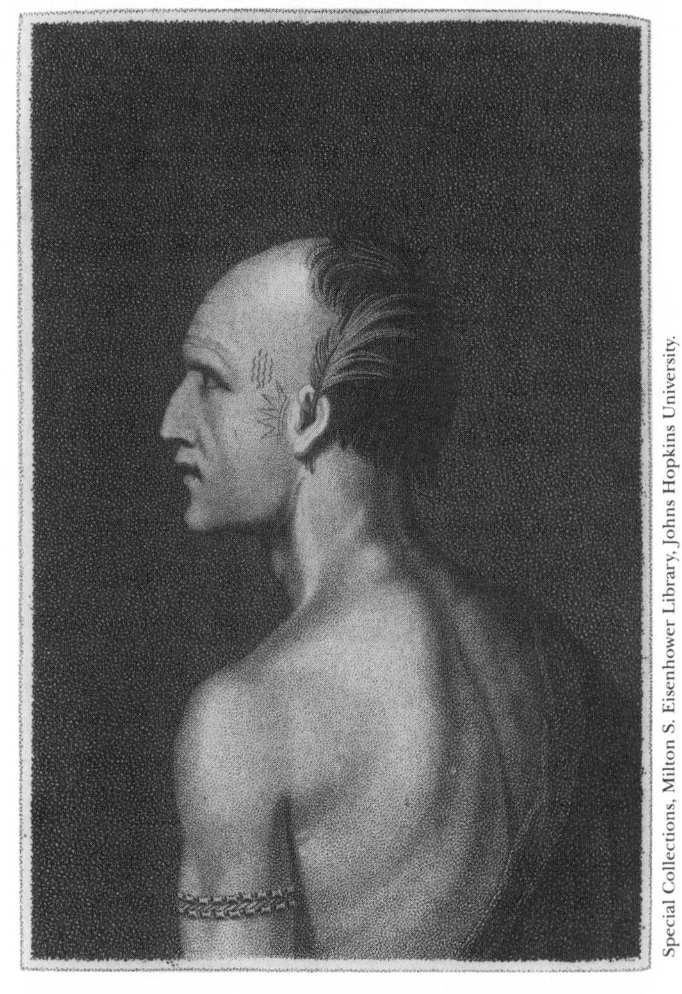

"Koohassen, Warrior of the Oneida Nation."

Crèvecoeur was adopted by the Oneidas.

In 1782 J. Hector St. John de Crèvecoeur asked, "What then is the American, this new man?" More than two centuries later, we are still trying to supply a satisfactory answer. So resonant, so fundamental is his question to contemplators of the American scene that its author's contemporary intellectual context has remained unexpounded. As in the works of William Bartram and Thomas Jefferson, natural history is the primary intellectual orientation of *Letters,* structuring the response to this question and governing the construction of the entire text. Seen in the light of natural history, it is possible to understand Crèvecoeur's choice of narrators, his long factual descriptions of various settlements in the American colonies, and narrator James's final decision to flee to the American Indians. More generally, natural history provides a context for the content of the entire book: *Letters* is an extended attempt to find and describe the type specimen of a new kind of man—the American.

Most critics of Crèvecoeur's *Letters from an American Farmer* have approached the book as a singular masterpiece. Largely unaware of the literary tradition from which it grew, they concentrate their efforts on a book that they regard as *sui generis.* Thomas Philbrick, however, recognizes the membership of *Letters* in "a . . . body of works which sought . . . to supply the British reading public with reliable accounts of the land and peoples of the troublesome North American colonies."[1] The tradition he correctly identifies and the works he names as the immediate predecessors of the *Letters*—including Carver's *Travels through the Interior Parts of North America* and Adair's *History of the American Indian*—contribute little to his analysis of Crèvecoeur's text. Long before Philbrick, D. H. Lawrence inaugurated the examination of *Letters* in isolation from any literary tradition, and most subsequent critics, including those who consider Crèvecoeur's indebtedness to the philosophical ideas of the French Enlightenment, follow this same procedure.[2] The book's formal and intellectual debt to the literature of place is rarely acknowledged and never analyzed.

In a 1785 letter to Thomas Jefferson, written from France three years after the publication of *Letters,* Crèvecoeur refers to lost portions of the manuscript, implying that his original aim in writing the book was to provide a comprehensive view of America, beyond the accounts he offers of Nantucket, Martha's Vineyard, Pennsylvania, and Charles Town: "I lost My Sketches of Maryland and so on Southerly. . . . Cou'd not you help me to Them in case of a Second Edition. Pardon the thought. It is not Vanity that Inspires it, but a desire that the Second Edition might be more usefull and more correct than the first."[3] In his desire to be "more correct" in a second edition, Crèvecoeur shares Jefferson's own motive in wishing to revise his *Notes*. In his plan to represent America comprehensively and in his desire to be of use Crèvecoeur allies himself with the writers of the literature of place. Use is uppermost, accompanied by a concern with encompassing the territory as completely as possible. In the parts of his manuscript that were preserved and published, the framework for Crèvecoeur's representation of America is natural history.

From the beginning Crèvecoeur makes clear the importance of this discipline to his text. The introduction provides character sketches of narrator James, the American farmer and of Mr. F. B., the fictional reader of the letters that the narrator will write. F. B. is an educated man. He has studied astronomy and mathematics. He has conducted James on a conversational tour of Europe, encompassing "navigation, agriculture, arts, manufactures, and trade."[4] Most significantly, he has studied natural history, and has engaged James to act as plant collector for him.[5] James and F. B. represent facets of Crèvecoeur's abilities and sensibilities. James is the simple farmer, competent in his practical agricultural domain but isolated on the frontier. F. B. is worldly, traveled, and educated. Although F. B. has studied a range of subjects, he actually practices botany. *Letters* has its genesis in a natural historian who enlists an American farmer to send him plants and write him letters about the new land.

Letters is a comprehensive look at America, constructed around a natural history core that takes characteristic forms of the manners-and-customs account and the natural history essay on a single kind of flora or fauna. Letters IV through IX are manners-and-customs accounts that describe the inhabitants of Nantucket, Martha's Vineyard, and Charles Town. Letter X, a natural history essay, describes New World snakes and hummingbirds. Letter XI is not natural historical in its method and rhetoric, but its subject is John Bartram, the internationally recognized colonial natural historian. This natural history core is framed by two narratives, one retrospective, one prospective, in farmer James's voice. After Letters I and II, James's initially vivid narrative presence fades during Letter III, to be replaced by a more objective, better educated, less distinctive voice that narrates the core of the book. In Letter XI, this remnant of James's voice is replaced by that of Iwan Al——z, the Russian gentleman who describes a visit to John Bartram. The James of the initial two letters returns for Letter XII, the apocalyptic account of his plan to flee with his family and join the American Indians. Natural history throws much-needed light on this narrator's seemingly dark plan to join the natives, and in turn, on the method and rhetoric of Crèvecoeur's James's most important contribution to American ideas: Letter III, "What is an American?"

Enough is known about Crèvecoeur's life and learning to create a convincing, if sketchy, portrait of him as a natural historian. Like Bartram, he was a traveler. Like Jefferson, he was a farmer. Crèvecoeur was born into the petty nobility in Caen, in the French province of Normandy, on 31 January 1735. He was educated first at home, and later by the Jesuits at the Collège Royal de Bourbon, where he attended as a boarding student. At the Collège he learned Latin, developed considerable skill in rhetorical exercises, and formed the habit of writing every day. He probably took a course in applied mathematics, in which he learned mapmaking and surveying. It is certain, at any rate, that he had those skills by the time he served in Canada as an engineer in the French Colonial Army.[6]

Crèvecoeur had traveled to Canada from England, where he had gone after his formal education to live with relatives.

The motivation for this move from France to England, the first of Crèvecoeur's immigrations, is not clear. His most recent biographers conjecture that he and his father argued over the younger man's choice of careers, his father intending him for the military, but Crèvecoeur refusing to enlist.[7] For whatever reason, after he finished school, perhaps as early as 1751, Crèvecoeur crossed the Channel to take up residence in Salisbury. There he perfected his English. Two years later he traveled to North America, where he served in the French Colonial Army in Canada as a surveyor and mapmaker. By 1759 he had arrived in New York, dropped there by a ship that took the rest of the men on board, French officers, back to France. Supporting himself with his surveying skills, Crèvecoeur traveled extensively among both white and American Indian peoples, and was adopted by the Oneidas.[8] On one of his journeys, undertaken in 1767, he accompanied Sir Robert Hooper on a surveying trip to the Indiana territory, traveling down the Mississippi to what is now St. Louis, and back to New York via the Illinois River, the Great Lakes, and the Hudson River—a distance of over 3,000 miles.[9] Surveying parties of this sort often augmented the activities of mapping and measurement with botanizing and collecting curiosities of natural history.

In 1769 Crèvecoeur married Mehetable Tippet and bought 250 acres of land in what is now Orange County, New York. This marks his second immigration. He farmed this land for about ten years, until his loyalist political inclinations made his further residence in America impossible. In 1779, he undertook his third immigration, fleeing with one son to France via New York, Ireland, and London, leaving his wife and two other children behind. He took with him a series of boxes, which were seized and searched when, ironically, the British authorities in New York City had him arrested as a suspected sympathizer with the Revolution.[10] The contents of these boxes were described by Peter Dubois, a magistrate assigned to interrogate Crèvecoeur: "When he came into this City, from among the Rebels, he brought with him Some Boxes in which he had curious Botanical plants and at the Bottom of those Boxes under the Earth in which these plants were, he had private Drawers or Cases in which he had his papers."[11] These papers, found at the bottom of the crates of plants that he was apparently taking to France, are widely suspected of being the manuscript of *Letters from an American Farmer.*[12] Crèvecoeur made it back to France by way of London, where he probably sold the manuscript of *Letters,* which was published in 1782.[13] After the American Revolution he returned to New York City as French consul, where he found his children safe, but his wife dead. He later sold his farm and relinquished his American citizenship, so as to be able to carry out his diplomatic duties without conflict of interest. Within two years he left his diplomatic post in America to return once again to France. He lived in his native country during her revolution, which notwithstanding his membership in the nobility, he survived. He died at Sarcelles, near Paris, in 1813. His obituary included the following summary of his activities during his long retirement: "Having held himself aloof from public responsibilities for over twenty years, M. de Crèvecoeur spent all his spare time spreading over France a taste for an improved agriculture and recommending, in numerous publications, most of them anonymous, the introduction of many useful plants and the tools best fitted to cultivate them."[14] Crèvecoeur the surveyor, mapmaker, traveler, and farmer was also Crèvecoeur the learned writer on agricultural subjects.

The evidence for Crèvecoeur's familiarity with natural history, particularly botany, is easier to assemble for the period of his life after his return to France, and after the publication of *Letters.* First, there is the box of plants itself. A man trying to leave the country, with a child in tow, chooses to box up plants to take along with him in his forced exile from his adopted home. Even if the plants were only camouflage for the precious papers at the bottoms of the boxes, he knew enough about the plant trade to choose this particular ruse. Magistrate Dubois, in the same document in which he described the contents of these boxes, noted among his prisoner's intellectual attainments, "Skilled in Botany."[15] While Crèvecoeur was farming in New York, he numbered Cadwallader Colden among his circle of friends.[16] Colden was a master of Linnaeus's method of classification and the author of *Plantae coldenghamae*, the first work on New York plants.[17] Just after his first return to France, in 1782—the same year that *Letters* first appeared—Crèvecoeur published under the name Normano-Americanus his *Traité de la culture des pommes de terre et des différents usages qu'en font les habitants des Étas-Unis de L'Amerique*. He thus anticipated by seven years the first publication of Parmentier, who is credited with the introduction of potato cultivation to France. In France Crèvecoeur's newfound friends, the philosophes of the French salons, took him to meet Buffon, whose works he had read and admired. During the course of his long French retirement, Crèvecoeur was honored by election to the Royal Agricultural Society and the Society of Agriculture, Sciences, and Arts at Meaux.[18]

Writing from New York in May 1785, Crèvecoeur sent the following thank-you to Thomas Jefferson, his one-time neighbor in New York, who was serving in Paris as minister to France: "I am much obliged to you for the directions you sent to Virginia. They have Proved of no avail. I have Received nothing. I wrote myself to richmond Last Fall. I have had no answer; that is both a loss and a disappointment. Luckily my Good Friend the vice consul of Carolina sent me an assortment Wherein I found the Magnolia Grandyflora and Do. umbrella."[19] This is one instance of Crèvecoeur's extensive correspondence on plants with a member of the natural history circle; other correspondents included Washington and Madison.[20] He knew the Bartrams as well.[21] The Colden-Jefferson-Bartram connection attached him to the American leaders of the international botanical circle. Botany and natural history were the logical pursuits of an intellectual farmer.

Crèvecoeur has a clear conception of the kind of book he is writing and is careful to distinguish his text from the standard European travel narratives. The minister, speaking in Letter I to James and his wife in his capacity as interpreter of Mr. F. B.'s instructions, speculates on the "modern travelers'" motive for choosing Italy as a destination: "I fancy their object is to trace the vestiges of a once-flourishing people now extinct. There they amuse themselves in viewing the ruins of temples and other buildings which have very little affinity with those of the present age and must therefore impart a knowledge which appears useless and trifling."[22] The narrator of Letter VIII also echoes this sentiment:

> Learned travellers, returned from seeing the painting and antiquities of Rome and Italy, still filled with the admiration and reverence they inspire, would hardly be persuaded that so contemptible a spot, which contains nothing remarkable but the genius and the industry of its inhabitants, could ever be an object worthy attention. But I, having never seen the beauties which Europe contains, cheerfully satisfy myself with attentively examining what my native country exhibits; if we have neither ancient amphitheatres, gilded palaces, nor elevated spires, we enjoy in our woods a substantial happiness which the wonders of art cannot communicate.[23]

Such passages help define the scope of *Letters* by creating a contrast with works such as Addison's *Remarks on Several Parts of Italy,* and with the practice of taking the grand tour. While he was in Salisbury, Crèvecoeur quite likely would have read Addison's enormously popular book, and he certainly would have learned of the custom of wealthy young Britishers "finishing" their educations with a tour through Europe. *Letters,* then, will not be a travel account in the Addisonian mode. It will not recount the country's cultural accomplishments for a learned audience. It will omit the fine arts.

Civil history is also to be ignored. Crèvecoeur's narrator in Letter III makes this clear in his account of Andrew the Hebridean:

> Let historians give the detail of our charters, the succession of our several governors and of their administrations, of our political

> struggles, and of the foundation of our towns; let annalists amuse themselves with collecting anecdotes of the establishment of our modern provinces: eagles soar high—I, a feebler bird, cheerfully content myself with skipping from bush to bush and living on insignificant insects. I am so habituated to draw all my food and pleasure from the surface of the earth which I till that I cannot, nor indeed am I able to, quit it.[24]

Letters will not do the work of annalists in recounting the country's civil history. After the fine arts and civil history have been eliminated, what remains, according to Thomas Jefferson's broad categories in his Baconian division of knowledge, is mathematical philosophy, moral philosophy, and natural history. Crèvecoeur, as an educated man, would likely have held approximately the same categories as Jefferson; Bacon's division was conventional at the time. Clearly, Crèvecoeur would not write a book on mathematics or physics. Nor would he write on ethics, although he is willing to judge as well as describe. His effort, *Letters,* is largely natural historical, concentrating on the subsection of natural history that Jefferson describes as the "occupations of man."[25]

Natural historical descriptions and subjects occupy the long middle of the book—Letters IV through XI. With the exception of Letter IX (on Charles Town) critics have little to say about these letters, after they have noted their exhibition of American simplicity, happiness, and prosperity.[26] Indeed, the letters' objectivity, their "logical coherence and system," does not invite the kind of interpretations available in the first three letters and in the final, seemingly apocalyptic, Letter XII.[27] The *Letters* reflected in the critics' interpretations is a book without a middle.

Of the eight middle letters, the five describing Nantucket and Martha's Vineyard are pure manners-and-customs accounts of the inhabitants of the two settlements. Near the beginning of Letter IV, Crèvecoeur's narrator once again specifies the kind of information he will include:

> I want not to record the annals of the island of Nantucket; its inhabitants have no annals, for they are not a race of warriors. My simple wish is to trace them throughout their progressive steps from their arrival here to this present hour; to inquire by what means they have raised themselves from the most humble, the most insignificant beginnings, to the ease and the wealth they now possess; and to give you some idea of their customs, religion, manners, policy, and mode of living.[28]

Letter IV, "Description of the Island of Nantucket, with the Manners, Customs, Policy, and Trade of the Inhabitants," is indeed a manners-and-customs account of the people, both white and American Indian, of Nantucket. It includes a description of the island's location: "The island of Nantucket lies in latitude 41° x 10'; 60 miles N.E. from Cape Cod; 27 N. from Hyannis, or Barnstable, a town on the most contiguous part of the great peninsula; 21 miles W. by N. from Cape Poge, on the vineyard; 50 W. by N. from Woods Hole, on Elizabeth Island; 80 miles N. from Boston; 120 from Rhode Island; 800 S. from Bermuda."[29] The incorrectness of this description notwithstanding, the method here is natural historical, as is the entire manners-and-customs account of the people living on the island. In using the method we have come to expect from a natural historian, and couching the resulting descriptions in the timeless, objective prose of the discipline, the narrator locates Nantucket in space, takes it consciously out of time—the rejection of civil history—and represents the island's land and people in a series of objective descriptions of the terrain, soil, products, and customs.

The narrator describes the island's plant productions: "it appears to be the uneven summit of a sandy submarine mountain, covered here and there with sorrel, grass, a few cedar bushes, and scrubby oaks; their swamps are much more valuable for the peat they contain than for the trifling pasture of their surface; those declining grounds which lead to the seashores abound with beach grass, a light fodder when cut and cured, but very good when fed green."[30] With all of the zeal—and the objectivity—of a modern-day ethnographer, the narrator describes the docks, the gardens, the community ropewalk, and the common meadow. He devotes two pages to the system of sheep-pasture titles by means of which the white islanders assigned the value of the land to its occupants, a system that created titles without necessitating subdivision.[31] He describes the method of fishing in the ponds, the Tètoukèmah lots for pasturing cattle, the land at the major settlements on the island, and the fish offshore. All of this description is couched in the impersonal historical present dictated by the rhetoric of natural history.

The customs of both white and American Indian inhabitants are described in the same passage, by means of the same rhetoric:

> The shores of this island abound with the soft-shelled, the hard-shelled, and the great sea clams, a most nutritious shell fish. Their sands, their shallows, are covered with them; they multiply so fast that they are a never-failing resource. These and the great variety of fish they catch constitute the principal food of the inhabitants. It was likewise that of the aborigines, whom the first settlers found here, the posterity of whom still live together in decent houses along the shores of Miacomet pond, on the south side of the island. They are an industrious, harmless race, as expert and as fond of a seafaring life as their fellow inhabitants, the whites.[32]

He speculates on the origin of these American Indians and, using a characteristic natural historical form, offers a list of the groups that have disappeared since the first white settlers arrived. He also accounts for the "peace and tranquillity" he found at Nantucket, once again describing the white inhabitants in the same rhetoric as the American Indian: "The simplicity of their manners shortens the catalogues of their wants; the law, at a distance, is ever ready to exert itself in the protection of those who stand in need of its assistance."[33]

This long description of the inhabitants of Nantucket,

couched in the objective, descriptive, atemporal rhetoric that Jefferson and Bartram used to describe their countries, is representative of the middle of the book. Letter V describes the "Customary Education and Employment of the Inhabitants of Nantucket," Letter VI is a "Description of the Island of Martha's Vineyard and of the Whale Fishery," VII returns to describe "Manners and Customs at Nantucket," and VIII, "Peculiar Customs at Nantucket," including the islanders' use of opium.[34] Beginning with Letter IV, and continuing through Letter VIII, the objective, orderly observation of the two seacoast settlements—from climate and terrain to people and their employments—is the main concern of the *Letters*.

In these middle sections the narrator, his identity as James dim but still recognizable behind the rhetoric of the natural historian, introduces a "they" in his description of the seacoast fisherfolk. James, of course, is the "I" of the book, his immediate neighbors, the "we." The narrator differentiates himself and his fellow inlanders from the inhabitants of the seacoast: "While we are clearing forests, making the face of Nature smile, draining marshes, cultivating wheat, and converting it into flour, they yearly skim from the surface of the sea riches equally necessary."[35] This separation of observer from observed is the now-familiar Othering of natural history. The domestic and economic arrangements of the seacoast dwellers might be described with affirmation and approval; nonetheless, they are the arrangements of strangers. This Othering will provide a model for reading the first three letters, particularly Letter III, "What is an American?"

Letter IX is also part of the natural historical core. "Description of Charles Town; Thoughts on Slavery; On Physical Evil; A Melancholy Scene" ends with a description of the slave in the cage. In this letter, critics usually detect a darkening and shift in the book.[36] The narrator signals this shift with warnings about the dangers of the heat: "The climate renders excesses of all kinds very dangerous, particularly those of the table; and yet, insensible or fearless of danger, they live on and enjoy a short and a merry life." He writes an essay on slavery, then returns to the subject of climate: "In the moments of our philanthropy, we often talk of an indulgent nature, a kind parent, who for the benefit of mankind has taken singular pains to vary the genra of plants, fruits, grain, and the different productions of the earth and has spread peculiar blessings in each climate."[37] The narrator launches on a "general review of human nature," examining the "vices and miseries peculiar to each latitude." It is an essay on climate's effects on nature, human and otherwise, and recalls the assertion in Jefferson's *Notes* of climate as the primum mobile of life. Here for the first time in *Letters*, the Great Chain has been invoked, for the idea of gradation and hierarchy is accompanied by a doctrine of plenitude, the belief that there are different kinds of creations, with different capacities, ordained to fill every niche in which a created thing might exist.[38] The narrator attributes his melancholy frame of mind to the scene with which the letter closes—the slave in the cage. He begins this section of the letter: "I was leisurely travelling along, attentively examining some peculiar plants which I had collected, when all at once I felt the air strongly agitated, though the day was perfectly calm and sultry."[39] This is the first reference to plant hunting since the introduction. Our narrator here possesses a natural historical habit that James, in the introduction, claimed he did not have. Recognizing plants as "peculiar" when not in one's home territory is an ability acquired through more than a casual, temporary acquaintance with the local flora. Crèvecoeur himself would have been capable of this discrimination. The description of the caged slave follows.

Letter IX is dark. The slave's condition, as the narrator renders it, is horrible. But the rhetoric is natural historical, with—as the references to plenitude and climate confirm—the same underlying assertions about order and cause and effect on which the five preceding letters were based. The "farmer of feelings," as Mr. F. B. calls James, has been made melancholy, but under this new emotion the objectivity of his observations is as unshaken as the objectivity of his observations about opium use on Nantucket, where those observations had a positive, approving overlay.[40]

This perspective on Letters IV through IX permits a better understanding of Letter X, "On Snakes: and on the Humming-Bird." Its form has gone unrecognized—it is two natural history essays, one on hummingbirds imbedded within an essay on snakes. The assignment has apparently come from Mr. F. B., and the narrator protests: "Why would you prescribe this task?" This is the first reference since his mention in the introductory letter of the agreed-upon practice of F. B. giving him his topics. He was capable of describing the manners and customs of the people on the seacoast and in Charles Town. He does not feel competent, however, to describe these animals: "You insist on my saying something about our snakes; and in relating what I know concerning them, were it not for two singularities, the one of which I saw and the other I received from an eyewitness, I should have but very little to observe."[41] The hummingbird, which lives only in the New World, and snakes with the supposed ability to mesmerize their prey are commonplaces of colonial natural history.[42] Ascribing emotion to the animals observed—"[hummingbirds] are the most irascible of the feathered tribe"—is also a commonplace.[43] The mode of presentation here is still natural historical and the rhetoric recognizably objective, for all of the anthropomorphic tendencies we can discern in the narrator's perceptions of these animals.

Given that the hummingbirds "fight with the fury of lions" and given that the snakes are observed "biting each other with the utmost rage," critics view this letter as a dark chapter in an increasingly darkening book. Lawrence praises it as "fine . . . in its primal, dark veracity." Elayne Antler Rapping concludes from this letter that the "state of nature is a state of war." Philbrick sees this letter as expressing nature's "deadly beauty," and he calls the narrator's detachment from the scenes he describes "a passive fascination strangely like that of the victims of the black snake."[44] These evaluations of the letter support the claim that the narrator's outlook gradually darkens as the book progresses. After the horror of the caged slave in the previous letter, an interpretation that shows nature red in tooth and claw is a logical one to make.

This, however, is a selective reading of the details. The narrator also describes a tame, defanged snake that acts like a pet cat: "[W]hen the boys to whom it belonged called it back, their summons was readily obeyed. . . . They often stroked it with a soft brush. . . . It would turn on its back to enjoy it, as a cat does before the fire." The description of the hummingbird also includes details that are rendered with approval: "On this little bird Nature has profusely lavished her most splendid colours; the most perfect azure, the most beautiful gold, the most dazzling red, are forever in contrast and help to embellish the plumes of his majestic head."[45] The portraits of these animals mix the violent with the beautiful, and even, in the case of the tame snake, with the domestic. Crèvecoeur might have intended a dark interpretation, but the details here, in contrast to those in the description of the caged slave, are mixed.

Letter IX has an unmistakable sentimental overlay: "I found myself suddenly arrested by the power of affright and terror; my nerves were convulsed; I trembled; I stood motionless, involuntarily contemplating the fate of this Negro in all its dismal latitude."[46] The narrator here becomes a participant in the scene he is describing. There is no corresponding engagement of the observer of the snakes and the hummingbirds. In Letter IX the narrator provides the dark conclusion in a mode characteristic of the sentimental, right down to the concluding "Adieu." In Letter X the critics provide the dark conclusion to descriptions that are mixed in their details, and whose natural historical rhetoric is unmediated by a moral or aesthetic overlay. In Letter X, then, the book returns from a dark view of the country to a more mixed, or neutral view of the land.

The natural history core culminates in Letter XI, "From Mr. Iw—n Al——z, a Russian Gentleman, Describing the Visit he Paid at My Request to Mr. John Bertram, the Celebrated Pennsylvanian Botanist." Most critics have stressed the peace and order of Bartram's farm, making little of the fact that the farmer is also a botanist. Rapping claims that Letter XI "describes the old age of an ideal American farmer." Philbrick acknowledges Bartram's botanical accomplishments, but finds that "first and foremost, Bartram is a farmer." Annette Kolodny sees this letter as representing "that same vision of pastoral harmony and noncompetitive social community that Farmer James had himself earlier invoked." David Robinson finds the Quaker farmer's freeing of his slaves to be an essential departure from James's practice in the first three letters.[47] If Bartram is included simply so that he might reassert the harmony and beauty of the freehold farm, then James himself could have narrated this letter. Crèvecoeur obviously felt that he must invent a narrator able to respond to Bartram's training as a systematic botanist and to appreciate his standing in the international botanical community. He substitutes the Russian traveler Iwan for the unlearned farmer James.

Iwan echoes James's belief in the effects of climate on a region's inhabitants, but the claim being made for the climate of Pennsylvania shifts slightly:

> In order to convince you that I have not bestowed undeserved praises in my former letters on this celebrated government, and that either nature or the climate seems to be more favourable here to the arts and sciences than to any other American province, let us together, agreeable to your desire, pay a visit to Mr. John Bertram, the first botanist in this new hemisphere. . . . It is to this simple man that America is indebted for several useful discoveries and the knowledge of many new plants. I had been greatly prepossessed in his favour by the extensive correspondence which I knew he held with the most eminent Scotch and French botanists; I knew also that he had been honoured with that of Queen Ulrica of Sweden.[48]

It is now the arts and sciences that will benefit from the good Pennsylvania climate, and Bartram is notable here not for his farming, but for his membership in the international botanical circle. This intellectual sophistication and his knowledge of American plants are what elicit his portrait in the *Letters*.

During Iwan's visit to Bartram's garden, discussion of botany occupies whole days: "Our walks and botanical observations engrossed so much of our time that the sun was almost down ere I thought of returning to Philadelphia; I regretted that the day had been so short, as I had not spent so rational an one for a long time before."[49] In addition to giving him a tour of the farm, Bartram tells Iwan how he came to study botany. Having become weary ploughing one day, he "ran under the shade of a tree" to rest. He plucked a daisy, and said to himself, "What a shame that thee shouldest have employed so many years in tilling the earth and destroying so many flowers and plants without being acquainted with their structures and their uses!"[50] He learned Latin, bought a copy of Linnaeus, and began to travel in search of botanical curiosities. He ends this account of his education and career in botany with an offer to send any trees or plants to Russia that Iwan might want. This narrative of his conversion to botany stands beside the other topics in Letter XI—the long description of Bartram's farm, the conversation about the treatment of his freed slaves, and the account of the Quaker meeting. Just as the description of the farm evokes the peace and happiness of James's own freehold described in Letter II, just as the description of the freed slaves contrasts with the slave in the cage described in Letter IX, just as the simplicity of the Quaker meeting echoes the numerous times throughout the *Letters* that Crèvecoeur's narrator praises the fairness and kindness of that sect, the choice of Bartram, of his educated European narrator, and the description of Bartram's botanical activity gather together and reemphasize the natural historical themes with which the core of the book has been concerned.

Botany is the centerpiece of natural history, and Bartram is the preeminent colonial botanist, one of the few to win worldwide recognition. He represents America's place in the grand, orderly scheme of nature. He is the man concerned with fitting America's productions into the Great Chain by means of the Linnaean system of description and nomenclature. He speaks a language that the rest of the world understands and respects. He contributes to a universal knowledge. He represents, in *Letters,* the perfect Quaker, the most enlightened of ex-slaveowners, the most prosperous Pennsylvania farmer, and at the same time, the renowned international botanical scholar. He is a figure of order in his family life, in his spiritual life, and in his intellectual life. Iwan, as an educated man, can read the letter from Queen Ulrica, can converse on botanical subjects with the botanist, and can appreciate the ingenuity of his farming methods and note the happiness of the ex-slaves, who have been freed to become wage earners. Iwan is Mr. F. B.'s counterpart, the only sort of narrator who is capable of describing a polymath like Bartram. Letter XI, then, is a representation of order on a grand as well as a local scale. It invokes nothing less than the worldwide order of systematic botany, of natural history, of the hierarchy of all created things.

Letter XI is the very opposite of dark. It celebrates reason, compassion, fairness, learning, and order. Natural history is the crowning order, with Quaker fairness, and Bartram's skill as a farmer providing echoes of the universal scheme in the religious community and in the family. Pennsylvania's climate permits this paradigm of all virtues to flourish. This same order, which Crèvecoeur's narrators endorse in their natural historical essays about the residents of Nantucket, Martha's Vineyard, and Charles Town, about New World snakes and hummingbirds, is reflected in the last letter of the volume.

In Letter XII, "Distresses of a Frontier Man," James returns as narrator. His assured state of mind, established in Letters I through III and sustained in a muted fashion throughout most of the natural historical core of the text, has been shattered by the onset of the Revolution. In James's analysis of the Revolution and the choices it forces him to make, politics and his family's welfare are opposed. Myra Jehlen points out that for James "civilization . . . was the family writ large." She notes that the law, government, king, governor, tax collector, were all most conducive to every virtue if they were remote from the citizens they ruled, governed, or taxed. Thus, Crèvecoeur's James's enthusiasm for King George is understood both as a function of the king's absolute authority to guarantee his subjects' rights, and by his extreme remoteness from those same subjects.[51] James knows he must preserve his family without angering either of the two factions—loyalists and rebels—who are waging war on the frontier:

> As a member of a large society which extends to many parts of the world, my connexion with it is too distant to be as strong as that which binds me to the inferior division in the midst of which I live. I am told that the great nation of which we are a part is just, wise, and free beyond any other on earth, within its own insular boundaries, but not always so to its distant conquests; I shall not repeat all I have heard because I cannot believe half of it. As a citizen of a smaller society, I find that any kind of opposition to its now prevailing sentiments immediately begets hatred; how easily do men pass from loving to hating and cursing one another! I am a lover of peace; what must I do?[52]

The "large society" was enlisting American Indians to burn the homes of settlers on the frontier. The "smaller society" was pressuring its members to support revolution. Those who did not were suspect. James defines the choice between the two factions—a political choice—apolitically: "I am conscious that I was happy before this unfortunate revolution. I feel that I am no longer so; therefore I regret the change."[53] The logic of James's decision to flee rests partly in his inability to side with either faction. The tribesmen in the hire of the loyalists would burn him out. The rebels, harassing anyone who was not a supporter of their cause, would drive him out. Letter XII is a virtual stream-of-consciousness account of the planning of his escape route as James arrives at a natural historical solution to his political problem: He will take his family into the wilderness to live in a native village. He has thought through every detail of this move. He will travel first by land for twenty-three miles, then by river. He will teach the villagers the value of farming, which he will continue to practice on the lands that the tribe allots him. His wife will learn to cook exotic foods such as corn and squash. He will build a wigwam.[54]

In the village he will reconstitute his family, the civilization of which political factionalism threatens to deprive him.

James's choice to run to the native Americans is often read as a move of absolute desperation. Philbrick, for example, calls it a "substitute for suicide."[55] It is understandable why James does not flee to Charles Town with its unpropitious climate and its resulting society of slaveholders, but why not go to the people of Nantucket or Martha's Vineyard? He approves of their society as heartily as he approved of his own farm. Perhaps they support the Revolution and would drive him out, just as his inland rebel neighbors threatened to do. A more fundamental reason for his decision to go west rather than east lies in the rhetoric of the letters that describe the seacoast. In those passages, the white settlers of Nantucket were as much Other as the American Indian islanders. That rhetoric compels us to see James's choice in Letter XII as a choice among possible Others. He chooses the interior-dwelling inhabitants because the circumstances of his present family situation are more closely approximated with them than with any of the other choices: unique among the people he has described, they farm in clearings they have made in the woods. They have little government, and little need for government. They will be a wholesome change from his politicized neighbors.

The logic of this thinking is accompanied by the panic and fear that lead critics to assert that this letter is apocalyptic.[56] James calls the change "this great transmutation," and "this proposed metamorphosis"; he imagines his family adopted by the native Americans and given new names.[57] This surrender of identity, of name, of outer shape, and even of the marks of species—a transmutation is the conversion of one species into another—suggests how essential this change would be. The natural history on which his observations of the Others has been based teaches the divine ordination of the divisions between kinds. That order is threatened by this transmutation that James envisions. This is an important source of James's ambivalence, and of the feverish tone of Letter XII: "however I respect the simple, the inoffensive society of these people in their villages, the strongest prejudices would make me abhor any alliance with them in blood, disagreeable no doubt to Nature's intentions, which have strongly divided us by so many indelible characters."[58] James simultaneously foresees his family's transmutation and plans to forestall it through providing his daughter with a husband of a suitable race. Yet he asserts, "Let us say what we will of them, of their inferior organs, of their want of bread, etc., they are as stout and well made as the Europeans."[59] This is a reference to the New World degeneracy hypothesis of the Abbé Raynal, to whom Crèvecoeur dedicates the *Letters*.

James worries, too, about his children's loyalties: "The apprehension lest my younger children should be caught by that singular charm, so dangerous at their tender years. . . . By what power does it come to pass that children who have been adopted when young among these people can never be prevailed on to readopt European manners?[60] James's concerns are racial and cultural. They stem from the categories imposed by natural history.

Letter XII, then—at first glance merely the feverish plan of a man driven to the edge by fear of being killed in the house he built with his own hands—presents a scheme that grows out of the natural historical method on which the entire book is based, and in whose rhetoric much of the book is cast. The native Americans might be Others, but so are most of the rest of the people in the *Letters*. "We" are few indeed, and can best be reconstituted among the American Indian "them." The solution, however, is not without its fearsome aspects. James's apprehensions are the fears of a man who is worried about his family turning into the Other. The confused, unsettled tone of Letter XII is not simply the product of a man who foresees changing his residence and way of life. Its cause is more fundamental. It is the product of a man who foresees changing his very identity, exchanging his own species for another. The darkness here is more than a foreshadowing of loss of country. It foreshadows loss of self.

Letter XII ends with a request for divine guidance. Like

Jefferson's passage in *Notes* in which he fears divine retribution for the treatment the white race has visited upon the blacks, Crèvecoeur's (James's) contemplation of the loss of his own racial identity prompts him to call upon the Supreme Being:

> If among the immense variety of planets, inhabited by thy creative power, thy paternal and omnipotent care deigns to extend to all the individuals they contain, if it be not beneath thy infinite dignity to cast thy eye on us wretched mortals, if my future felicity is not contrary to the necessary effects of those secret causes which thou hast appointed, receive the supplications of a man to whom in thy kindness thou has given a wife and an offspring; view us all with benignity, sanctify this strong conflict of regrets, wishes, and other natural passions; guide our steps through these unknown paths and bless our future mode of life.[61]

The reference to the immense variety of inhabited planets is to the doctrine of the "plurality of worlds," a central feature of which was the belief that these worlds were inhabited.[62] This doctrine is one expression of the Supreme Being's plenitude, the order of which is expressed by the Great Chain. James invokes the Supreme Being, who has ordained that order so completely represented by John Bartram in the previous letter. He is careful to request only that felicity for himself and his family that does not conflict with science, with the "necessary effects of those secret causes which [the Supreme Being] has appointed." Natural history suggested that the Other was a different species. But it also suggested that the society of this particular Other, whose climate, employment, and apolitical inclinations were so like his own, was the best place to "be restored to [his] ancient tranquillity."[63] There is hope, in this concluding prayer, that transmutation or no, order will prevail. The final letter is a fitting ending to the long middle of *Letters*. In it the natural history that has governed and generated the book threatens James with the loss of his identity, while holding the promise of an orderly outcome to his chaotic situation.

When James decides to take his family to live among the American Indians, he not only abandons politics ("Not a word of politics shall cloud our simple conversation") but also takes himself out of history.[64] He will become one of the native Americans, one of those static figures in the tableau of natural history who have only origin and existence, who are without history. But he was not so different from them even before the political troubles intervened. In Letter II, "Situations, Feelings, and Pleasures of an American Farmer," James narrates a personal history that he shares with a great many other immigrants and that is essentially without connection to civil history. He is the living, breathing, talking representative of the type that the book is trying to define—the American.

This identification of James as representative of the true American is reaffirmed in Letter III by means of the method of natural history, whereby an observer nominates an individual to be a type specimen. Several possible types of the true American are considered and rejected as Crèvecoeur's James struggles to answer the question, "What then is the American, this new man?"[65] At the beginning of the letter he assumes the point of view of the traveler, concluding for him that "we are the most perfect society now existing in the world." For the purposes of the introduction to the letter, the Americans are, with few exceptions, "all tillers of the earth."[66] Climate, however, will change various groups of newly formed Americans: "The Americans were once scattered all over Europe; here they are incorporated into one of the finest systems of population which has ever appeared, and which will hereafter become distinct by the power of the different climates they inhabit." As he essays a survey of these differences, he tells us "[m]en are like plants; the goodness and flavour of the fruit proceeds from the peculiar soil and exposition in which they grow."[67] He briefly mentions sea dwellers, then claims that frontier dwellers "near the great woods . . . grow up a mongrel breed."[68] However, he approves of "a second and better class, the true American freeholders." Twice he selects "an epitome" of such a person. First he relates the history of an unnamed "he," then the story of the final type specimen, Andrew the Hebridean. "What is an American?" then, receives a series of answers as the narrator sorts through various peoples to select the type specimen from a locale where "the air we breathe, the climate we inhabit, the government we obey, the system of religion we profess, and the nature of our employment" combine to produce a true American type.[69] Despite the inclusive generalization at the start of the letter, where the narrator claims that an immigrant "becomes an American by being received in the broad lap of our great Alma Mater," the epitome, the type, is finally Andrew the Hebridean, who transforms himself from a penniless immigrant into a freehold farmer.

This description of the American type reached in the relative calm of Letter III helps explain James's feverish reasoning in Letter XII. To describe this type specimen, James must recite that man's individual history, for it is as much in his history as in his eventual status as a freeholder that his Americanness resides. Yet that history is not seen as a unique set of events that happens to an individual and that distinguishes him from all others. It is seen as a representative set of events, as typical of the history of all others of the same type. Andrew the Hebridean has a history in the same sense that an individual plant has a history. It changes over time in that it grows to become a representative of its type. This is an instance of the flattening of narrative that the texts of Bartram and Jefferson exhibited. For Crèvecoeur's James, the result is a type of an American taken out of time, out of history.

When James describes the "type" of American in the manner of a natural historical specimen, by implication he takes all Americans out of history, too. Like plants, they exist in certain climates and assume certain characteristics because of their environments: "It is with men as it is with the plants and animals that grow and live in the forests; they are entirely different from those that live in the plains."[70] They are the natural produce of the countryside; they are not the product of historical circumstance, nor do they, themselves, make history, aside from their unremarkable personal histories. In Letter XII, when history more momentous than the anonymous history of individuals threatens to be made, James wants no part of it.

At stake in Letter XII is the identity not only of an individual but of the type of the American as well. Like the type specimen Andrew the Hebridean, James is an inland freehold farmer. His grandfather shared the essentials of Andrew's history—the hard work to clear the land, to build a house and barn, to leave a freehold to his children. Most important, James has in common with the Americans the climate, the soil, and his employment. Thus, by the logic of the natural history that governs *Letters*, James is an American. Staying at his farm and siding with the rebels is ultimately as threatening to his identity as an American as fleeing to the natives on the frontier. Both options involve a change in circumstances significant enough to endanger the type. For Crèvecoeur and his narrator, the type of the American presented in *Letters* is tenuously maintained, defined only for a moment, and at a given place. As the conditions that lend identity to Americans change, conditions such as climate and government, the type changes. People who were once Americans find themselves in a new climate, or the subjects of a new government and the possessors of a new identity. Ultimately, then, the progression of the book is from a description of the American, and his embodiment in James the freehold farmer, to a signaled but not yet completed change in that description. The new definition of the term "American" must await the effects of the climate and government in James's new, aboriginal neighborhood, where he will repeat his grandfather's personal history. The darkness of Letter XII is the shadow of change passing over the static conception of the American sanctioned by the method and rhetoric of natural history. Letter XII is not a substitute for suicide. It is the desperate attempt of a man to maintain his intellectual equilibrium when the very basis of order in his cosmos is threatened.

For Crèvecoeur, colonial America is not, finally, a melting pot, his most famous image notwithstanding. It is a place in which, for a brief period, government could be kept remote and civil history forestalled, a place in which a true American type, the inland freehold farmer, could flourish. The continued existence of that type is threatened by civil history, by politics, and by the unknown influence of the aboriginal Other. James's flight is undertaken to preserve the purity of the type. For all of its projector's fevered calculation and its wild destination, it is, from James's point of view, a profoundly conservative change. He will preserve the climate, he will recover the lax government that suits him, he will take with him his religion and will not have to change his employment. His wife and children will go along, and with them he will reconstitute his family, the basis, for him, of civilization. He will trust the order of Nature.

Crèvecoeur leaves us with an image of an American defined by the land, not by the government or by historical events. The laws of natural history—which is to say, the laws of Nature—permit his being, and these laws are threatened by civil history and politics. Thus, Crèvecoeur leaves us with an image of an American fleeing into the ungoverned wilderness, into a place that has only natural history, to preserve his Americanness. Crèvecoeur's American is defined outside of politics and prior to civil history, in atemporal natural historical conditions in which, like a plant, he can flourish. America for Crèvecoeur is the land, not its government. On the possibility of reconciling the two, he is silent. He sees only as far as the reconstitution in the wilderness of the conditions that gave rise to the true American.

D.H. Lawrence accuses Crèvecoeur of bad faith in his decision to return to France when the impending American Revolution made untenable his continued residence at his New York farm: "[It] was all a swindle. Crèvecoeur was off to France in high-heeled shoes and embroidered waistcoat, to pose as a literary man, and to prosper in the world."[71] Lawrence thought that Crèvecoeur should have done what his narrator had planned—he should have gone to live with the American Indians. It is possible that it was as much of an adjustment for Hector St. John, as Crèvecoeur called himself in America, to remember how to be a Frenchman as it would have been for him to learn how to be an American Indian. By the time he returned to France, Crèvecoeur had forgotten much of his French.[72] He had lived in America for most of his adult life. He had traveled among the native Americans, traded with them, and been adopted by them. He had collected their legends and had even written one in their style, deftly reflecting their worldview. This legend, "The Cherokee Tradition," is a transformation tale in which a miserable man and the members of his family are "by a sudden metamorphosis" changed into beavers.[73] Crèvecoeur himself was adept at transformation, but he paid the price that James foresaw, the dissolution of the self as it changes from American to Other. After his move from America to France, Crèvecoeur was almost consumed by anxiety over his children, legal problems in claiming his inheritance, and fear for his own safety during the French Revolution. History had intervened in Crèvecoeur's own life, spoiling the atemporal serenity of his existence as an American farmer.

Albert E. Stone claims that Crèvecoeur's *Letters* inaugurates American literature "as the voice of our national consciousness," if by literature one means "arrangements of affective images embodied in the traditional forms of poetry, fiction, and drama, and expressing the spirit of place."[74] Crèvecoeur's James spends part of what some critics have called a "romance" inhabiting the romance's setting—America.[75] Letters I and XII are particularly vivid accounts of moments in James's life. Much of the rest of the time, the work defines and creates that setting and the people who live in it through physical description of the land itself, and through exhaustive portraits of the inhabitants of that land. The scene in which American literature is set was not, for Crèvecoeur, a given in the way it was for the British novels of the same period.

This difference can be illustrated by matching James the farmer of feelings against his British counterpart in the novel named for him, *The Man of Feeling* (1771). The brief introduction of the British novel assumes a geographical setting, then quickly sketches a cultural and literary setting as well. A curate and a landholder are shooting "on a piece of fallow-ground." For wadding, the curate is using the manuscript of a novel. The landholder rescues what turns out to be *The Man of Feeling* from the curate and presents it to the reader.[76] Geography, the place of the two men in the culture, the business of the hunt, the literary tradition in which the manuscript properly belongs ("had the name of a Marmontel, or a Richardson, been on the title-page—'tis odds that I should have wept")—all are indicated with the economy that only tradition permits. The "spirit of place" in the British work is in each detail, and it is in the reader, who shares the context into which to put those details. It was the task of Crèvecoeur to construct that context for his reader. In America, there was no "there" there, and there was not yet a "they" there either. Natural history, not the distinctive individual narrative of the actions of individuals, was the means to construct that context.

Like the other representatives of the literature of place, *Letters* presents a portrait of America. It is a land of happy freehold farmers and whalers and fishermen, of dissipated planters and miserable slaves. It is a land of disappearing American Indians. James turns America's inhabitants into natural history types and arrays them in a landscape waiting for civil history. James is the only one of these type specimens who talks, breathes, doubts, or thinks in anything approaching a conventional novelistic representation. When history arrives, James imagines a flight into the wilderness, to the Other, to distance himself from the Others he has taken such pains to describe. The story of his flight inaugurates American literature, because James, at the end of the work, has what he lacked at the beginning—a setting that is clearly defined in the reader's mind. Letter XII, finally, is James's declaration of independence from the political situation that has engulfed him, from the history that is about to be made, and, although he could not know this, from what would turn out to be not Others, but Americans. Paradoxically, as *Letters* inaugurates American literature, James divests himself of the claim to be the first American novelistic character, just as Crèvecoeur makes a similar divestiture of his claim to be the first American novelist. In the end, neither James nor Crèvecoeur is a postrevolutionary American. They both become Other. And the scene of the first work of American literature is left once again a static tableau, awaiting both history and its fictional counterpart, the novel.

• CHAPTER FIVE •

The Passing of Natural History and the Literature of Place

Courtesy of the Pennsylvania Academy of Fine Arts, Philadelphia. Gift of Mrs. Sarah Harrison (the Joseph Harrison, Jr., Collection)

Charles Willson Peale, The Artist in His Museum *(1822).*

Charles Willson Peale invites us into his museum of natural history.

Travel literature has remained a flourishing, if minor, genre. Despite the invention of photography, motion pictures, and videotape recording there is still an audience for the verbal description of foreign places and peoples. Gone, however, is the political urgency that during the years of the American Revolution elevated certain instances of colonial American travel literature into the literature of place. Gone, too, is the foundation of natural history that supplied a universal context in the absence of a national one.

Like Addison, the contemporary American travel writer carries his cultural and political contexts with him. The description of a foreign land reflects the cultural preconceptions of the writer and is simultaneously an exploration of the unfamiliar and a reaffirmation of the home culture. By contrast Bartram the failed Floridian, Jefferson the Virginian, and Crèvecoeur the American farmer wrote as insiders, as inhabitants (or one-time inhabitants) of the territories they described. These men composed their books during the revolutionary period—after the passage of the Stamp Act in 1765 and before the federal government was established in 1789. The declaration of hostilities against Great Britain had removed the old colonial framework. No substitute was supplied until the establishment of the federal government. They lacked a political context for their works, situating them instead within the the universal, scientific framework of natural history.

Written in 1784, just one year after Great Britain recognized American independence, John Filson's *Discovery, Settlement, and Present State of Kentucke* reflects both the growing influence of the newly independent American states and the waning influence of natural history. Beginning in 1783, Filson, a schoolmaster, made at least four trips from his native Pennsylvania to the Kentucky territory. On 19 December 1783 the surveyor of Fayette County, Kentucky, recorded in Filson's name 12,368½ acres of land located in three separate tracts. In an effort to increase the value of this land, Filson researched and wrote *Kentucke* to serve as an inducement to additional settlers. The research consisted of traveling through the wilderness and of interviewing a handful of pioneers who knew Kentucky well—including Daniel Boone. By 1784, Filson had completed his manuscript and returned to Wilmington, Delaware, to have it printed. Filson's final western trip ended in 1788 when he was killed on the banks of the Great Miami River, where he had gone to survey the site of present-day Cincinnati. He was thirty-five years old.[1]

Like the works of Bartram, Jefferson, and Crèvecoeur, Filson's *Kentucke* is a description of a territory. Unlike these men, Filson writes as an American, for Americans. The dispute that prompted William Bartram in his *Travels* to substitute dashes for the name of the *Prince of Wales,* the brigantine on which he sailed from Philadelphia to Charleston; the fighting that drove Jefferson out of Monticello during the writing of his *Notes;* the increasingly menacing political situation that sent Crèvecoeur and his narrator, James, fleeing their farms in opposite directions had no similar effect on Filson. The danger for him was not from the British, but from the native Americans who were fighting for the frontier. Unlike Bartram, Jefferson, and Crèvecoeur, Filson could begin to rely on a new national context for his representation of Kentucky. He could describe a territory that existed in relation to a series of adjoining, independent states. This point of view is reflected in an assumption Filson makes in writing *Kentucke*—the new settlers he seeks to attract will come from American territory. He concludes his book with a table of distances from cities in the settled areas of America, such as Pittsburgh and Philadelphia, to various points in Kentucky.[2] For Filson, as for many writers who would follow him, travel to the western wilderness is defined as travel away from established states, and after 1789, away from an established country. Unexplored western territory could be represented using America itself as a context.

Despite the new American political context for the description of Kentucky, Filson employs the conventions of natural history in *Kentucke*. He casts a cursory glance at the soil, plants, and animals of the territory. But the authorization for Filson's representation of Kentucky is not the cosmic, teleological order implied by natural history. Instead, it is an appendix of *Kentucke* entitled "The Adventures of Col. Daniel Boon; Containing a Narrative of the Wars of Kentucke," which is the first written account of the frontiersman's story and the only part of Filson's book that has endured. Natural history in *Kentucke* is made possible by Boone's story. Human agency creates the conditions of possibility for what was once a divinely teleological Great Chain and system of nature.

The organization of the text roughly follows the now-familiar natural historical pattern. The title page promises "An Essay towards the Topography and Natural History" of Kentucky, and after a brief history of the territory's "Discovery, Purchase, and Settlement," Filson describes the area's natural history in the next ten sections: "Situation and Boundaries," "Rivers," "Nature of the Soil," "Air and Climate," "Soil and Produce," "Quadrupeds," "[White] Inhabitants," "Curiosities," "Rights of Land," and "Trade of Kentucke." Compared with Bartram's or Jefferson's accounts of natural history, Filson's is sketchy. Quadrupeds, for example, are dispensed with in three short paragraphs.[3]

Nonetheless, Filson uses the rhetoric of natural history. He describes a single plant: "Here is a great plenty of fine cane, on which the cattle feed, and grow fat. This plant in general grows from three to twelve feet high, of a hard substance, with joints at eight or ten inches distance along the stalk, from which proceed leaves resembling those of the willow." This description permits a reader to reconstruct the appearance of sugarcane. He makes lists: "The waters have plenty of beavers, otters, minks, and musk-rats: Nor are the animals common to other parts wanting, such as foxes, rabbits, squirrels, racoons, ground-hogs, pole-cats, and oppossums."[4]

In these early sections of *Kentucke,* then, we see the natural history pattern arrayed—the divisions sanctioned by earlier texts employing the same rhetoric, the lists used by readers to supply links on the Great Chain. The portion of the text that employs this natural historical rhetoric and method also manifests the atemporal, ahistorical nature of that rhetoric. Filson was aware of the topics conventionally discussed in natural history essays. The resulting representation is of a Kentucky that is a static, atemporal tableau vivante.

Civil and personal history also figure in Filson's description of Kentucky. The story of Daniel Boone is the most important of these narratives. In the texts of Bartram, Jefferson, and Crèvecoeur, narrative remained subordinate to natural history. As we have seen, it was a competing mode of discourse. Time, the essence of narrative, was at odds with the atemporal eternality implied by natural historical description and sanctioned by the Great Chain. In Filson, the meeting of this natural historical rhetoric with the competing and antithetical rhetoric of narrative results once again in the removal of the American Indians to their own section of the text (which foreshadows their eventual removal from Kentucky as well). But this text is unlike the others in that this removal is not, ultimately, made possible by the assumptions of natural history. In *Kentucke,* the assumption that native Americans are describable using the same method and rhetoric that a natural historian might use to describe a plant proves inadequate to the task of subduing the natives long enough to describe them using that rhetoric. Filson must look elsewhere to find a sanction for his description of the American Indians, and by implication, for all of Kentucky. He finds this sanction in the Daniel Boone narrative.

Kentucke begins with a short essay on civil history, an account of European attempts to achieve clear title to land in Kentucky. This brief civil history is left inconclusive: "[N]otwithstanding the valuable Considerations [the American Indians] received, [they] have continued ever since troublesome neighbours to the new settlers."[5] The longest narrative, the Boone appendix, is not couched in the rhetoric of this opening section—the rhetoric of civil history—but as personal history, told from the first-person point of view. "The Adventures of Col. Daniel Boon" interrupts the natural historical description, coming after the sections on "Curiosities," "Rights of Land," and "Trade," but before an account of the Pinkashaw Council and the manners-and-customs description of Kentucky's native Americans. Filson's Boone sets the scene of his narrative in a way that owes nothing to the conventions of natural history:

> We had passed through a great forest, on which stood myriads of trees, some gay with blossoms, other rich with fruits. Nature was here a series of wonders, and a fund of delight. Here she displayed her ingenuity and industry in a variety of flowers and fruits, beautifully coloured, elegantly shaped, and charmingly flavoured; and we were diverted with innumerable animals presenting themselves perpetually to our view.[6]

This is an aesthetic response to the wilderness. The setting in which Boone enacts the events in his narrative is not the representation of the land afforded by the natural historical method and rhetoric. At the point in Filson's text that Boone acts, the natural historical setting is incomplete. The American Indians have not yet been described, and it is their behavior toward the whites that defines Boone's reason for being in Filson's text. Instead of a natural historical setting, which is unavailable to him, Boone acts in a setting represented by a purely aesthetic approach to landscape. And there is little enough of even this Boone-as-aesthete. The narrative concentrates on episode, not description. Boone encounters a problem, prevails against it, and moves on to the next problem.

Early in his account Boone is captured by the American Indians. Later he is captured again. Much of the intervening narrative describes a series of battles with the natives—Boone narrates at least a dozen separate engagements, carefully recording the number of whites wounded and killed in each clash. His second captivity takes place in lieu of a battle. He surrenders when he and his men are ambushed during a trip away from the garrison.[7] During this captivity he is adopted by his captors: "At Chelicothe I spent my time as comfortably as I could expect; was adopted, accordin[g] to their custom, into a family where I became a son, and had a great share in the affection of my new parents, brothers, sisters, and friends."

Boone's adoption takes place just as the settlers' war with the natives is turning. The American Indians, outnumbered by newly reinforced settlers, and "out-generalled in almost every battle" begin to avoid formal warfare and "practice secret mischief."[8] Boone's second capture is a clear example of differing warfare strategies.

Filson's Boone supplies a curious counterpart to the description of America's native inhabitants in natural historical texts. Instead of providing objective, natural historical information about American Indians, Boone becomes one. As their adopted son, Filson's Boone assumes the manners and observes the customs of the people who capture him. The reader sees Boone in his American-Indian family instead of in his white one. Day-to-day life in frontier white society is excluded from the narrative: "Soon after this, I went into the settlement, and nothing worthy of a place in this account passed in my affairs for some time." Long journeys with his white family are similarly passed over: "The history of my going home, and returning with my family forms a series of difficulties, an account of which would swell a volume, and being foreign to my purpose, I shall purposely omit them."[9] The depiction of white society in the garrisons and in Boone's family gives way to first-person accounts of battles and enforced sojourns with the Indians. The first-person point of view of Boone's narrative is crucial. It eliminates the Othering inherent in the third person, and, of course, it individualizes Boone's experiences. The reader hears, in the voice of the person experiencing it, an account of a white settler on the frontier. This is the only white man in the entire book who speaks as a white man.[10] He does this while acting like an American Indian.

In the introductory history of Kentucky, Filson introduces Boone, but as we have seen he leaves the American Indian troubles unresolved. Boone resolves them in his narrative: "Many dark and sleepless nights have I been a companion for owls, separated from the chearful society of men, scorched by the Summer's sun, and pinched by the Winter's cold, an instrument ordained to settle the wilderness. But now the scene is changed: Peace crowns the sylvan shade."[11] The problem with which the book began is now solved. Kentucky is safe for white settlement.

Boone's narrative ends with his defeat of the American Indians. The next section of the text, the account of the Pinkashaw Council, offers the reader a chance to hear the native Americans themselves accept peace terms. The section on the council is followed by the manners-and-customs account, "Of the Indians."[12] No longer a threat, no longer an active element in the settlers' lives, the native Americans can be reduced through the rhetoric of natural history to an aggregate Other, their tribes and nations listed, their numbers counted, their origins from the Welsh asserted, their bodies, "genius," and religion described.

Boone has made Kentucky safe for white settlement, and *Kentucke* safe for natural history, for the scientific removal of American Indians from the rest of the text. In Boone's narrative, Kentucky's civil history is completed and its natural history sanctioned. He puts an end to the hostilities that kept Kentucky's civil history incomplete while at the same time rendering the native Americans fit subjects for natural history.

The standard manners-and-customs account, which Filson supplies after the Boone narrative, is, like all such accounts, atemporal. It denies American Indians any history, either personal or cultural, and places them in a static tableau. Until they could be subdued, that is, until they could be counted on not to step forward out of that tableau brandishing a scalping knife, their natural history was unavailable, the natural historian at such risk that observation and writing were impossible. Of course, Boone did not literally provide protection for Filson as the schoolmaster gathered natural historical "discoveries" and observed the American Indians, although the manner of Filson's death suggests that he needed someone like Boone to keep him safe. In the text, however, Boone provides the authority that natural historical description of a hostile people requires.

Boone does not prevail against the native Americans until

they adopt him. He becomes Other himself in order to secure the American Indian as Other. Like the manners-and-customs description that represents native Americans in Filson's narrative, Boone's narrative is also separate. Soon after publication, the Boone narrative was excerpted from Filson's *Kentucke,* becoming much more famous in its own right than the parent text.[13] As Richard Slotkin has explained, Boone's stature became mythic, and the myth became the type for a long line of American heroes—Cooper's Natty Bumppo, Wister's Virginian, Melville's Ahab.[14] Boone's mythic transformation takes him out of history just as surely as the American Indians' natural historical description takes them out of history. The conclusion to *Kentucke,* a vision of a prosperous Kentucky peopled by Europeans, reflects Boone's mythic function:

> In your country, like the land of promise, flowing with milk and honey, a land of brooks of water, of fountains and depths, that spring out of valleys and hills, a land of wheat and barley, and all kinds of fruits, you shall eat bread without scarceness, and not lack any thing in it; where you are neither chilled with the cold of capricorn, nor scorched with the burning heat of cancer; the mildness of your air so great, that you neither feel the effects of infectious fogs, nor pestilential vapours.

This biblical benediction is concluded by an extraordinary sentence: "Thus, your country, favoured with the smiles of heaven, will probably be inhabited by the first people the world ever knew."[15] Filson here addresses the residents of the Kentucky for whom Boone has prepared the ground. Boone himself is the Ur-Kentuckian, and his narrative is simply one more explanation of origins: the American Indians were descended from the Welsh; Kentuckians, from Boone. And they are a new race, "the first people." Thus, in the text Boone is a figure intermediate between chaotic savagery and the civilization that is to come. He creates the historical certainty necessary for a new race to emerge on the once-Dark-and-Bloody Ground.

Perhaps Filson subscribed to the natural historical theory

that there were more races of men than *Homo sapiens*. In this view, the adopted Boone has been transmutated—he has joined the Other. The very act that enabled his elimination of the American Indians as a factor in Kentucky's history also disqualified him from membership in Filson's new race of "first people." There is no place for an "Indian" like Boone in the Kentucky that Filson foresaw. Boone is a civilizing force but is not himself civilized. Ultimately, he is confined in his own section of the text, separated like those who adopted him from the promised land that he helped make safe for the new people yet to come.

In Filson's text, the sanction for the natural historical description of a territory shifted from the eternal, universal, divine order of the Great Chain to the temporal, local, human order achieved through violent subjugation. The natural historical descriptions thus sanctioned begin, in Filson, to seem perfunctory. They are sketchy, concentrating on the sensational, ignoring other more commonplace, but no less exotic, natural productions. Filson can be careless about providing at least the beginnings of a complete record of Kentucky's natural productions because he understands that Boone is the real authority for his representation of Kentucky. In contrast to the natural historical record, the Boone narrative is complete, each battle listed, the casualties noted. The complete authority for the representation of place is no longer the creator of the Great Chain. Human agency provides the safe conditions in which science can furnish representations of a given country.

Filson's *Kentucke* marks the passing of natural history in another way as well. In addition to bolstering divine authority for natural history with the military victories of an adopted son of the American Indians, he also includes a glimpse of the changing view of the world, which toppled certain key assumptions of natural history and precipitated its transformation into the life sciences. In 1766 at Big Bone Lick, a salt spring near the Ohio River, mammoth bones had been found. These bones counted as a natural production of Kentucky, and Filson devoted a few pages of his text to their description, including in that description a conjecture as to how the mammoth had become extinct. His belief in the extinction of species, although shared by such luminaries as Buffon, was far from universal. Thomas Jefferson fully expected Lewis and Clark to find evidence that the mammoth was still alive. His view of the creation did not admit the possibility of an end to a species. Filson takes the opposite view, speculating that the American Indians had hunted "the tyrant of the forests, perhaps the devourer of man" to extinction. He understands the contradiction he has introduced into the doctrine of the Great Chain: "Can then so great a link have perished from the chain of nature? Happy we that it has. . . . To this circumstance [the native Americans' hunting] we are probably indebted for a fact, which is perhaps singular in its kind, the extinction of a whole race of animals from the system of nature.[16] Filson hedges. Aware of the challenge that extinction presents to the traditional, atemporal system of nature, he asserts that the mammoths' extinction might be "singular."

Other thinkers were not so hesitant. Georges Cuvier (1769–1832), a pioneering animal anatomist and reformer of the system of animal nomenclature, also classified fossils, and he realized that fossil animals had become extinct. Although he was not an evolutionist, his theory of catastrophism, which explained extinction through postulating a series of deluges, inserted into the study of living things a crucial element it had lacked—time. In 1858 this temporality would become installed permanently as the accepted explanation for fossil animals and for variation within living species when Charles Darwin and Alfred Wallace published in the *Journal of the Linnaean Society* an explanation of their theory of evolution by natural selection. The atemporality that had permitted Bartram, Jefferson, and Crèvecoeur to describe a landscape believed to be static was replaced by a temporality that stretched for aeons and encompassed all natural things.

Cuvier's influence on natural history was far-reaching. Foucault presents his achievement in a startling image: "One day, towards the end of the eighteenth century, Cuvier was to topple the glass jars of the Museum, smash them open and dissect all the forms of animal visibility that the Classical age had preserved in them."[17] Cuvier's accomplishment was his insistence on looking beneath the surface of things. Natural history as Bartram, Jefferson, and Crèvecoeur practiced it was a science of surfaces. In 1814, in a letter protesting innovations in the system of nomenclature put forward by Cuvier and Jussieu (who changed the basis of plant classification in his *Genera plantarum)*, Jefferson described the boundaries of the discipline of natural history:

> However much . . . we are indebted to both these naturalists, and to Cuvier especially, for the valuable additions they have made to the sciences of nature, I cannot say they have rendered her a service in this attempt to innovate in the settled nomenclature of her productions; on the contrary, I think it will be a check on the progress of science. . . . Their systems . . . are liable to the objection of giving too much into the province of anatomy. It may be said, indeed, that anatomy is a part of natural history. . . . But . . . it has been necessary to draw arbitrary lines, in order to accomodate our limited views. According to these, as soon as the structure of any natural production is destroyed by art, it ceases to be a subject of natural history, and enters into the domain ascribed to chemistry, to pharmacy, to anatomy, etc.[18]

Cuvier's contribution was his destruction, by art, of the structure of natural productions, particularly animals. He is the father of comparative anatomy. He performed dissections and classified organs and organ systems according to their function. He could reconstruct, from a single bone, the structure of an animal because he understood the necessary relationship of one part of an animal's body to all others. As Foucault points out, Cuvier shifted the means by which relationships were established in the science of life:

> [I]n contrast with the mere gaze, which by scanning organisms in their wholeness sees unfolding before it the teeming profusion of their differences, anatomy, by really cutting up bodies into

> patterns, by dividing them up into distinct portions, by fragmenting them in space, discloses the great resemblances that would otherwise have remained invisible; it reconstitutes the unities that underlie the great dispersion of visible differences. The creation of the vast taxonomic unities (classes and orders) in the seventeenth and eighteenth centuries was a problem of *linguistic patterning*: a name had to be found that would be both general and justified; now, it is a matter of an *anatomic disarticulation;* the major functional system has to be isolated; it is now the real divisions of anatomy that will make it possible to form the great families of living beings.[19]

In his letter, Jefferson seeks to redraw a boundary that Cuvier had blurred when he sought in "the organism's inner darkness" the structures beneath the surface that establish similarities with other organisms. The order implied by the great effort at description and classification that grew out of the Linnaean method—an order based on visible, external characteristics—was being replaced by functional characteristics. The position of a living thing in the grand order was also shifting: "Before, the living being was a locality of natural classification; now, the fact of being classifiable is a property of the living being."[20] That is, the living being had become more than simply an entry on the great table of living things in which its identity was embodied by its classification. Classification was moved off center stage. Function assumed its place. Biology was born.[21]

The effort to classify all of the natural productions of the world continues. Botanists still make collecting trips, the specimens they find are still sent to herbaria, and binomial names are assigned following the rules in the *International Code of Botanical Nomenclature*. There are similar procedures for animals. The utility of a single name for each natural production is still recognized. What has shifted is the explanatory power with which that name has been invested by its users. That shift severs the connection between the table of names and teleology that Bartram, Jefferson, and Crèvecoeur could invoke as a context for their descriptions of America.

One branch of travel literature has developed and expanded into its own discipline—ethnography. Its practitioners face many of the same questions and issues that Bartram, Jefferson, and Crèvecoeur did in their efforts to create verbal representations of people of other cultures. There is as yet no Cuvier or Darwin to free the texts of ethnographers from the flawed and partial representation of other cultures that has come down to them from the earliest travel accounts through natural history. And the assumptions of ethnographers—the Othering, the atemporality, the individual- and history-effacing generality of ethnographic descriptions—are very much a part of contemporary thinking.

At the National Museum of Natural History in Washington, D.C., in the series of exhibits devoted to America's native peoples, is a display case depicting the "Polar Eskimo."[22] Against a painted background of an igloo built on an ice shelf there are five dark-haired, brown-skinned mannequins dressed in fur parkas and mukluks—three adults and two children. Standing with them are five stuffed sled dogs. The object of attention of both humans and canines is a small stuffed seal lying next to a hole painted on the bottom of the floor of the display case. The seal is attached to the hands of one of the adults by a fishing line. The caption on the interpretive plaque reads, "Call that a seal?" The display is a reconstruction of the kind of manners-and-customs descriptions that Bartram, Jefferson, and Crèvecoeur wrote: static, frozen, concerned with the surfaces of people—their clothing, racial characteristics, dwellings, and diet.

The caption on the plaque prompts the viewer to ask three sets of related questions. The first set has to do with the accuracy of the caption, which seems to assert that anyone who sees a smaller rather than a larger dinner on the end of a fishing line would have such a response. But is irony a characteristic response of the Inuits to a situation of this sort? And who would have been the speaker? Would a wife have said such a thing to her husband? Would a child have said such a thing to a parent? A second set of questions concerns the very idea of a "characteristic response." How can the characteristicness of a response be determined? How can the nomination of any such response as characteristic constitute anything other than a cultural stereotype? A third set of questions involves the alternatives to this caption. Why not record what an actual Inuit said under just such a circumstance? This would be analogous to Jefferson's quoting the Mingo chief Logan in Section VI of his *Notes*. But, given that this is an exhibit on *the* Polar Eskimo, rather than *a* Polar Eskimo, such a caption would imply a characteristicness that the accidental utterance of an individual might not have. The implication that any utterance used as a caption will be taken as characteristic returns the questioning back to the first set. The representation of the "characteristic" is the problem here, as it was in Bartram, Jefferson, and Crèvecoeur.

The first museum exhibitions of the type still on display at the National Museum of Natural History were mounted in Philadelphia during the period that Bartram, Jefferson, and Crèvecoeur wrote. Charles Willson Peale (1741–1827) announced plans for his museum of natural history on 7 July 1786.[23] Peale was a successful portrait painter and had assembled a series of his paintings into a Gallery of Great Men that included Franklin, Jefferson, and eventually, William Bartram.[24] He expanded the portrait gallery to accommodate natural historical specimens, which he became expert at preserving and mounting. His skill as an artist led to the display of specimens in cases with backdrops painted to represent each specimen's natural habitat, a very early instance of such a technique. Peale's museum was not the first publicly displayed collection of natural history in America. In 1773, the Library Society of Charleston had organized an exhibit of items representing South Carolina's natural history.[25] The exhibit burned in 1778, and the war prevented its replacement.[26] But just twenty-seven years after Sir Hans Sloane's collections were bought by Parliament and exhibited in London under the name "British Museum," Peale was planning to open a similar institution in Philadelphia.

Peale included in his museum a number of mannequins with wax faces: "a Chinese 'Laborer and Gentleman,' the Indian sachems Blue Jacket and Red Cloud, a Hawaiian, a Society Islander, and a few others, attired in donated clothing and accouterments."[27] The method of display on view in the National Museum of Natural History today is essentially the one pioneered in America by Charles Willson Peale in the 1780s.

Questions of representativeness, of the ethnographic text as text and the ethnographer as writer, are now being asked by anthropologists. Stephen A. Tyler identifies the rhetoric of natural history as a source of problems in ethnographic texts and discovers consequences of the sort we have seen in the three writers examined in this study:

> The urge to conform to the canons of scientific rhetoric has made the easy realism of natural history the dominant mode of ethnographic prose, but it has been an illusory realism, promoting, on the one hand, the absurdity of 'describing' nonentities such as 'culture' or 'society' as if they were fully observable, though somewhat ungainly, bugs, and on the other, the equally ridiculous behaviorist pretense of 'describing' repetitive patterns of action in isolation from the discourse that actors use in constituting and situating their action, and all in simpleminded surety that the observers' grounding discourse is itself an objective form sufficient to the task of describing acts.[28]

Less bitterly, but making the same point, Clifford Geertz observes that anthropologists are unaccustomed to asking "[h]ow words attach to the world, texts to experience, works to lives."[29] The relatively new practice of "look[ing] *at* as well as through" anthropological texts he attributes to a realization of the new political situation in which ethnographers now find their subjects:

> The transformation, partly juridical, partly ideological, partly real, of the people anthropologists mostly write about, from colonial subject to sovereign citizens, has . . . altered entirely the moral context within which the ethnographical act takes place. . . . One

> of the major assumptions upon which anthropological writing rested until only yesterday, that its subjects and its audience were not only separable but morally disconnected, that the first were to be described but not addressed, the second informed but not implicated, has fairly well dissolved.[30]

The natural historical rhetoric that Bartram, Jefferson, and Crèvecoeur employed was sanctioned in their day by the idea of the Great Chain. The order of the Chain assured them, in their depiction of American Indians, that they were superior to native Americans, and that the method and rhetoric created an appropriate image of native peoples. Crèvecoeur, it will be remembered, also turned the method and rhetoric on the white inhabitants of America. Geertz's observation about the susceptibility of colonial peoples to unexamined ethnological description would apply to Crèvecoeur as well as to contemporary ethnologists. They found colonials, including white colonials, susceptible to natural historical ethnological treatment. Looking at, rather than through, ethnographic texts and considering ethnographers as writers, Geertz claims, should result in a more artful writing: "If there is any way to counter the conception of ethnography as an iniquitous act or an unplayable game, it would seem to involve owning up to the fact that . . . it is a work of the imagination."[31] Crèvecoeur found this to be so. His largely ethnographical work shades off (on both ends, as it happens) into prose fiction. If, as Geertz implies, there is an affinity between art and ethnography, if ethnography is a work of the imagination, then prose fiction, the genre that seeks to depict individuals in universal situations, is on one end of a continuum with ethnography, the genre that seeks to find universal situations in individuals on the other end. Bartram and Jefferson cluster at the ethnography end of this continuum; Crèvecoeur is nearer to prose fiction.

When Crèvecoeur's James describes members of his own culture using the same rhetoric that he used in representing the Other, he anticipates one of the newer developments in ethnography: the Other employing the dominant culture's media, forms, and rhetoric to represent itself to an audience composed of members of the dominant culture. On display near the "Polar Eskimo" diorama in the National Museum of Natural History are three totem poles. Built into the pedestal on which they rest is a television monitor on which a video depicting the lives of the Kwakiutl people runs in a continuous loop. The Kwakiutl, who carved these soaring sculptures, live in the Pacific Northwest. They are perhaps best known for the custom of the potlatch, a feast at which the host, who is celebrating a special event such as the naming or the marriage of a child, feeds his neighbors and bestows lavish gifts upon them. At the beginning of the video, Gloria Cranmer Webster, a Kwakiutl and museum curator says, "When you look at museum exhibits in a lot of places, it is as if we were gone." The video stresses the continuity of Kwakiutl tradition in contemporary times. In one shot, taken at a potlatch, we see a man wearing a mask and cape, arms outstretched, stepping a syncopated circle around a fire while the community watches. In a second shot, we see a young boy preparing to learn a ceremonial dance. An adult dresses him in a headdress and a cape. The music begins, he stretches out his arms, and steps the same shifting rhythm that the older dancer did. Under the cape the boy wears a pair of shorts and a tee shirt. The elders who accompany him on musical instruments display a similar contrast in their dress. The static atemporality of the display case is replaced, in the video, by the moving child dressed in both traditional and contemporary clothing. In the video, the Kwakiutl represent themselves as a living people who maintain a continuity with their past. The narrative representation of the Other that eluded Bartram, Jefferson, and Crèvecoeur is finally being supplied by the Other themselves.

Henry David Thoreau, writing more than fifty years after the three major authors considered here, contrasts his effort in *Walden* (1854) with accounts of distant peoples in distant lands: "I would fain say something, not so much concerning the Chinese and Sandwich Islanders as you who read these pages, who are said to live in New England. . . . I have travelled a good deal in Concord."[32] In this early passage from the first chapter of *Walden,* Thoreau is explaining to his reader the sort of book he is writing. He distinguishes his text from travelers' accounts of native peoples in far-off places (Thoreau's ethnographic subjects are his own neighbors) and, pretending not to know what he knows very well, adopts the careful locution of the self-consciously objective observer ("who are said to live in New England"). Thoreau's topics in *Walden* coincide with those in the literature of place—the land itself, its plants, animals, and inhabitants. However, Thoreau's procedure for observing and recording his observations retains but a remnant of the natural historical method that Bartram, Jefferson, and Crèvecoeur employed.

That method had as its basis visual inspection followed by a written account of the natural object being studied. At the heart of the written account was the Linnaean system of nomenclature, a universal set of names and descriptors that permitted initiates to the method to converse with one another without ambiguity about any natural production. It also permitted the location of any natural production on the Great Chain of Being. A Linnaean observer either collected a specimen or located one already in a collection, observed its surface features carefully, and wrote a universally recognizable description of that object. The method itself dictated the surface features that were to receive this scrutiny. A plant would be described by its flower, fruit, leaves, and seeds; these elements were considered essential. All others, however interesting, were accidental and could safely be ignored. The method emphasized the discovery of similarities among individuals in order to place them in the same group. Differences between a given group and all other groups divided each species from the next. As we have seen, the natural historical practice of Bartram, Jefferson, and Crèvecoeur was largely circumscribed by this method. It suffused their thinking, provided them with categories, and dictated their prose styles.

Thoreau's tone in the early passage from *Walden* quoted above warns us that in his text the conventions of natural historical writing are subject to ironic manipulation. When critics read *Walden* with belletristic assumptions, that is, when they find in *Walden* thematic and symbolic patterns typical of a work of belles lettres, they are not falling into the kind of anachronistic misreading that we have seen such assumptions lead to in various interpretations of Bartram, Jefferson, and Crèvecoeur. Thoreau's *Walden* is an example of natural historical writing that is not scientific, but meditative, in contrast to Bartram's *Travels*, Jefferson's *Notes*, and Crèvecoeur's *Letters*, which could, and did, contain both passages that are observations of external nature and passages that are the reports of the observers' reactions to external nature. These latter we are tempted to label merely meditative, but the writers themselves, like Edmund Burke, would have seen such exclamations as scientific. By the time Thoreau wrote *Walden*, the scientific and the meditative no longer inhabited the same discourse. For the writers of scientific texts in the mid-nineteenth century, the outlines of the natural world (which had been traditionally described as the Creation) were subject to change over time. Even taxonomy changed. Establishing the affinity of one group of natural productions with another became a more complicated procedure than simply observing and comparing surface characteristics. As we have seen, the method of biology shifted away from observation of surfaces to dissection. In the late eighteenth century, at about the same time that Cuvier was reintroducing dissection into Western science, Joseph Priestley conducted the experiments that resulted in the identification of the gases that are consumed and produced during photosynthesis. At the time Thoreau wrote *Walden*, observations of the visible surface components of plants and animals were still important in the process of assigning names to natural things, but these observations did not comprise the whole of taxonomy, and taxonomy itself was not the whole of botany or biology.

In *Walden*, Thoreau is often content to observe the surface of natural productions. He loosely follows many of the forms laid down by natural history, but his goal is literary, not scientific. In his chapter "Brute Neighbors," he writes a series of essays whose textual forebears are the natural history essays of Bartram and Crèvecoeur. Thoreau had read Bartram; in *Walden* he quotes the *Travels* at length.[33] His chapter on the woodland animals shares its form with passages in the *Travels* in which Bartram strings together a series of natural history essays.[34] Thoreau's subjects are a "wild, native" mouse, the phoebe, robins, the partridge, the woodcock, ants, and the loon. For the mouse, partridge, and loon he provides binomials. Thoreau prefaces these essays by two questions that contrast his worldview with the assumptions of natural history: "Why do precisely these objects which we behold make a world? Why has man just these species of animals for his neighbors; as if nothing but a mouse could have filled this crevice?"[35] For Bartram, Jefferson, and Crèvecoeur, the idea of the Great Chain precluded these questions. They mark the distance between a metaphysics that can seem arbitrary and Alexander Pope's typically eighteenth-century assertion, "One truth is clear, 'Whatever is, is right.'"[36]

Perhaps the shift from the eternal, static gradation of the Great Chain to a more easily questioned metaphysics narrowed the range of observations that were considered scientific. When science in the form of natural history seemed so clearly to reflect the order of the created world and the mind of the Creator, wonder and awe in the face of that creation were an appropriate subject for a scientific text, as long as these emotions were securely grounded in empirical observation. Thus we have Bartram's frequent descriptions of the sublime scenes in the American Southeast. In the new, less-fixed context, where time and change hold sway, a narrowing of subjects accompanies the inevitable rise in skepticism that the new context engenders. When things change, an observer can be certain of less, and a careful observer will constrict the circle of his observations. Thoreau, then, locates *Walden* outside of the boundaries of the scientific discourse of his day on two counts—his observations are of the old, surface, natural historical type, and his reactions to the natural world are not excluded from his text. Natural history, for Thoreau, is an anchor for the meditative side of *Walden*. The information he supplies is undoubtedly accurate, but it is no longer the sort of information that a scientific audience finds useful.

He is perfectly aware of the scientific inutility of his writing. In his essay on mice, he begins by noting that "the mice which haunted my house were not the common ones, which are said to have been introduced into the country," adding, "I sent one to a distinguished naturalist, and it interested him much." We do not learn why the mouse interested the naturalist, but we do learn why it interested Thoreau—it is the occasion for his asking the metaphysical question that begins the series of brief essays ("as if nothing but a mouse could have filled this crevice"). He provides us with a carefully observed narrative of the behavior the mouse exhibited after he had coaxed it out of hiding with the crumbs of his lunch ("it . . . cleaned its face and paws, like a fly").[37] For the naturalist, the mouse was strange in the scientific sense; it was a new animal to describe. For Thoreau, the mouse was strange in a metaphysical sense; its existence (rather than some other being's) seemed arbitrary and became the occasion for a meditation on what is. The meditation precludes science, which had narrowed its focus as it began to hang its observations on a much less certain, much more plastic worldview.

To end his chapter on "Brute Neighbors," Thoreau describes his pursuit, in a canoe, of a loon as again and again the bird dives, swims underwater, and resurfaces. The motion of the submarine bird is impossible to predict; he "manœuvred so cunningly that [Thoreau] could not get within half a dozen rods of him."[38] The unseen movement of the bird under the water, Thoreau's futile efforts to anticipate the bird's surfacings and surprise him by paddling to that spot—the narrative of these actions attracts the sort of interpretation that Bartram, Jefferson, and Crèvecoeur's texts, based as they are on an unironic application of natural historical principles, ought not to. The unfathomable subsurface movement of *Columbus glacialis* might be an analogue to the new science of interiors and dissection in which Thoreau does not participate. He is the natural historian, still looking at the surface of things. To know the loon as a contemporary scientist knows him requires subsurface observation. This subsurface knowing—a knowledge of organs and structure, of bodily organization—does not interest Thoreau, who is concerned with the meaning in creatures that comes from their souls, "the *anima* in animals."[39]

Observation of surfaces, the scientific rhetoric of the natural historians, functions in *Walden* as an anchor for Thoreau's meditations, his speculations on the *anima*. As a superseded scientific discourse, it retains its objectivity without compelling the writer employing it to fulfill every requirement of its once strictly enforced method. Thoreau feels no obligation to describe the mouse, birds, or ants completely, or to notice structures, even those on the surface, that are essential to scientific discourse. Because Thoreau could choose his subjects, and then make a further selection of the elements of those subjects that he would include in *Walden*, interpretations of nature-as-symbol are available to the critic.

For Bartram, Jefferson, and Crèvecoeur the rhetoric and method of natural history provided a context and a framework for the representation of a country. In their books, these authors' reactions to the land, couched in a discourse that we are now tempted to regard as purely literary, stand beside their observations on natural productions, couched in a discourse that we can only with effort recognize as scientific. After the passing of natural history, this once-scientific discourse was converted to literary purposes.[40] In the hands of Thoreau and others like him it became the vehicle for a representation of America that could not only provide a description of her natural productions but also convey her symbolic meanings.

Notes

PROLOGUE: RECOVERING A LOST PARADIGM

1. J. Hector St. John de Crèvecoeur, *Letters from an American Farmer and Sketches of 18th-Century America,* ed. Albert E. Stone (New York: Penguin, 1963) 71. Crèvecoeur, like all eighteenth-century naturalists, takes the male as representative of the species.

2. Crèvecoeur 7.

CHAPTER ONE. NATURAL HISTORY IN CONTEXT

1. Moses Coit Tyler, *The Literary History of the American Revolution 1763–1783,* 2 vols. (New York: G. P. Putnam, 1898) 2: 10.

2. Bernard Chevignard, "St. John de Crèvecoeur in the Looking Glass: *Letters from An American Farmer* and the Making of a Man of Letters," *Early American Literature* 19 (1984): 180–81. Chevignard casts doubt on the traditional dates given for the composition of the *Letters*. He argues that Crèvecoeur did not have a completed manuscript of the book when he fled his Orange County farm in 1779.

3. Russel B. Nye citing Samuel Miller, *The Cultural Life of the New Nation, 1776–1830* (New York: Harper and Brothers, 1960) 93.

4. Nye 87.

5. William Darlington, *Memorials of John Bartram and Humphrey Marshall* (Philadelphia: Lindsay and Blackiston, 1849) 342.

6. Chevignard 174.

7. Brooke Hindle, *The Pursuit of Science in Revolutionary America, 1735–1799* (Chapel Hill: Univ. of North Carolina Press, 1956) 13.

8. Edmund Berkeley and Dorothy Smith Berkeley, *The Life and Travels of John Bartram* (Tallahassee: Univ. Presses of Florida, 1982) 18.

9. Berkeley and Berkeley 19; Ernest Earnest, *John and William Bartram: Botanists and Explorers* (Philadelphia: Univ. of Pennsylvania

Press, 1940) 39. However, see David Scofield Wilson for a harsher judgment of Collinson's relationship to Bartram: *In the Presence of Nature* (Amherst: Univ. of Massachusetts Press, 1978) 95.

10. Berkeley and Berkeley 19.

11. Berkeley and Berkeley 36.

12. William Thomas Stearn, "Notes on the Illustrations," Appendix, *Species Plantarum: A Facsimile of the First Edition 1753,* by Carl Linnaeus, 2 vols. (London: The Ray Society, 1959) 2: 63.

13. Darlington 97.

14. Darlington 152.

15. Darlington 104.

16. Earnest 41.

17. Hindle 8.

18. Joseph Ewan, ed., *William Bartram: Botanical and Zoological Drawings, 1756–1788* (Philadelphia: American Philosophical Society, 1968) 164.

19. Arthur O. Lovejoy, "The Place of Linnaeus in the History of Science," *Popular Science Monthly* 71 (1907): 500.

20. Arthur O. Lovejoy, *The Great Chain of Being* (Cambridge, Mass.: Harvard Univ. Press, 1973) 232.

21. In building my argument using Foucault's observations on Linnaeus and natural history, I do not wish to be understood as ascribing to the larger intellectual program that he presents. His local observations about certain texts, audiences, and their interactions can be separated from his statements about language and its role in human institutions. His readings of natural historical texts that embody the Linnaean method provide provocative descriptions of how the texts were likely read by their eighteenth-century audience. His insistence on excavating the ground of possibility of knowledge—the *episteme*—is a radical instance of a reader approaching older texts on their own terms. It is possible to benefit from these readings without invoking his larger program. This is what I do.

22. Michel Foucault, *The Order of Things* (New York: Vintage, 1973) 142.

23. John Bartram, "Diary of a Journey through the Carolinas, Georgia, and Florida from July 1, 1765 to April 10, 1766," *Transactions of the American Philosophical Society* ns 33 (1942): 31.

24. William Bartram, *Travels,* ed. Mark van Doren (New York: Dover, 1955) 369.

25. Ewan 151.

26. William Bartram, *Travels* 369n.

27. John Reinhold Forster, "Account of Several Quadrupeds from Hudson's Bay," *Selected Works by Eighteenth-Century Naturalists and Travellers,* Introduction by Keir B. Sterling (New York: Arno, 1974) 370.

28. Forster 370.

29. Forster 374.

30. Forster 1.

31. Ewan 163.

32. William Bartram, *Travels* 324–25.

33. Ewan 155.

34. William Thomas Stearn, "Sources, Format, Method and Language of the *Species Plantarum,*" *Species Plantarum: A Facsimile of the First Edition 1753,* by Carl Linnaeus, 2 vols. (London: The Ray Society, 1957) 1: 82–83.

35. Carl Linnaeus, *Species Plantarum: A Facsimile of the First Edition 1753,* 2 vols. (London: The Ray Society, 1959) 2:938.

36. Modern usage would dictate that the two-word specific epithet be hyphenated (*noli-tangere*), thus making the binomial name (*Impatiens noli-tangere*) literally two words, not three.

37. Stearn, "Sources, Format" 1:94.

38. While it is true that the species name of *Impatiens noli-tangere* means "not touching," the verb form is not a predicate. It has been reduced to a noun.

39. Sten Lindroth, "The Two Faces of Linnaeus," *Linnaeus: The Man and His Work,* ed. Tore Frängsmyr (Berkeley: Univ. of California Press, 1983) 157, 163, 176.

40. Steven Jay Gould, *The Mismeasure of Man* (New York: W. W. Norton, 1981) 35, 40–41.

41. Gunnar Broberg, "*Homo sapiens:* Linnaeus's Classification of Man," *Linnaeus: The Man and His Work,* ed. Tore Frängsmyr (Berkeley: Univ. of California Press, 1983) 157.

42. Mary Louise Pratt, "Scratches on the Face of the Country; or, What Mr. Barrow Saw in the Land of the Bushmen," *Critical Inquiry* 12 (1985): 120, 121, 127.

43. Foucault 150.

44. For another view of the atemporal nature of these narratives, see Wayne Franklin, *Discoverers, Explorers, Settlers: The Diligent Writers of Early America* (Chicago: Univ. of Chicago Press, 1979) 21. For a suggestive account of American natural historians engaging in "taxonomic construction as a rehearsal . . . of social and political

construction" see Christopher Looby, "The Constitution of Nature: Taxonomy and Politics in Jefferson, Peale, and Bartram," *Early American Literature* 22 (1987): 252–73.

45. Although Foucault is at pains to claim that the temporal, represented by evolutionism, which was considered by some eighteenth-century thinkers, and the atemporal, whose apotheosis is Linnaeus's system, were "complementary" (150), the very fact that he identifies the two ideas and admits that time and the system are not integrated, permits a consideration of them as separate entities.

46. Charles L. Batten, Jr., *Pleasurable Instruction* (Berkeley: Univ. of California Press, 1978) 36, 42–43, 46, 10.

47. Captain Harry Gordon, "Journal of Captain Harry Gordon," *Travel in the American Colonies, 1690–1783,* ed. Newton D. Mereness (New York: Macmillan, 1916) 455–89.

48. Bishop John Frederick Reichel, "Travel Diary of Bishop and Mrs. Reichel and their Company," *Travel in the American Colonies, 1690–1783,* ed. Newton D. Mereness (New York: Macmillan, 1916) 583–99.

49. Moses Coit Tyler groups these texts under the heading "Descriptions of Nature and Man in the American Wilderness: 1763–1775." 1: 141.

50. Patricia M. Medeiros, "Three Travelers: Carver, Bartram, and Woolman," *American Literature 1764–1789: The Revolutionary Years,* ed. Everett Emerson (Madison: Univ. of Wisconsin Press, 1977) 195.

51. Robert Rogers, *Journals of Major Robert Rogers* (1765; n.p.: Readex Microprint, 1966) v.

52. Robert Rogers, *A Concise Account of North America* (1765; New York: Johnson Reprint Corporation, 1966) vii.

53. William Smith, *Historical Account of Bouquet's Expedition against the Ohio Indians in 1764* (Cincinnati: Robert Clarke Company, 1907) 23.

54. William Stork, *Account of East Florida* (London: n.p., 1766) Dedication, i.

55. James Adair, *Adair's History of the American Indians,* ed. Samuel Cole Williams (New York: Promontory Press, 1930) xxxv, xxxvi.

56. Jonathan Carver, *Travels through the Interior Parts of North America in the Years 1766, 1767, and 1768,* 3rd ed. (London: n.p., 1781) xiv–xv.

57. William Bartram, *Travels*, ed. Mark van Doren (New York: Dover, 1955) 15.

58. Julian P. Boyd, ed., *The Papers of Thomas Jefferson*, 20+ vols. (Princeton: Princeton Univ. Press, 1950–) 8: 155.

59. Rogers, *Journals* 205.

60. John R. Stilgoe, "Fair Fields and Blasted Rock: American Land Classification Systems and Landscape Aesthetics," *American Studies* 22 (1981): 23.

61. Rogers, *Journals* 66, 95.

62. Rogers, *Journals* 100.

63. Rogers, *Concise Account* 90–124.

64. Rogers, *Concise Account* 264.

65. Smith 62–63.

66. Smith 63, 64, 66.

67. Stork 40–41.

68. Pratt 120.

69. Adair xxxv.

70. Adair 428.

71. Annette Kolodny, "Review of *Narratives of North American Indian Captivities*," *Early American Literature* 2 (Fall 1979): 228.

72. Richard Slotkin, *Regeneration through Violence: The Mythology of the American Frontier, 1600–1860* (Middletown, Conn.: Wesleyan Univ. Press, 1973) 96; Roy Harvey Pearce, *The Savages of America: A Study of the Indian and the Idea of Civilization* (Baltimore: Johns Hopkins Press, 1965) 43; Louise K. Barnett, *The Ignoble Savage: American Literary Racism, 1790–1890* (Westport, Conn.: Greenwood Press, 1975) 26.

73. Pearce 142–43.

74. Carver 414–41.

CHAPTER TWO. *DESCRIPTION AND NARRATION IN BARTRAM'S* TRAVELS

1. William Bartram, *Travels* (Philadelphia: James and Johnson, 1791).

2. John Hendley Barnhart, "Bartram Bibliography," *Bartonia* 12 (1931): 51–2.

3. Lane Cooper, "A Glance at Wordsworth's Reading," *Modern Language Notes* 22 (April 1907): 110–17; John Livingston Lowes, *The Road to Xanadu* (Boston and New York: Houghton Mifflin, 1930); N.

Bryllion Fagin, *William Bartram: Interpreter of the American Landscape* (Baltimore: Johns Hopkins Univ. Press, 1933).

4. Robert D. Arner, "Pastoral Patterns in Bartram's *Travels*," *Tennessee Studies in Literature* 18 (1973): 133–46; Wayne Franklin, *Discoverers, Explorers, Settlers: The Diligent Writers of Early America* (Chicago and London: Univ of Chicago Press, 1979); Thomas Vance Barnett, "William Bartram and the Age of Sensibility," diss. Georgia State Univ., 1982, 49, 52.

5. Barnett 50; Fagin 6–9.

6. Joseph Ewan, ed. *William Bartram: Botanical and Zoological Drawings, 1756–1788* (Philadelphia: American Philosophical Society, 1968) 6.

7. Edmund Berkeley and Dorothy Smith Berkeley, *The Life and Travels of John Bartram* (Talahassee: Univ. Presses of Florida, 1982) 278.

8. Brooke Hindle, *The Pursuit of Science in Revolutionary America, 1735–1799* (Chapel Hill: Univ. of North Carolina Press, 1965) 14.

9. Ewan, *Drawings* 6.

10. Hindle 13, 27.

11. William Darlington, *Memorials of John Bartram and Humphrey Marshall* (Philadelphia: Lindsay and Blackiston, 1849) 344.

12. Darlington 347–48.

13. Darlington 346.

14. The only mention of the *Travels* before 1786, when Enoch Story, Jr., proposed to publish it, was by Johann David Schoef, who saw the manuscript in 1783. *Travels: Naturalist's Edition,* ed. Francis Harper (New Haven: Yale Univ. Press, 1958) xxi.

15. Linnaeus called John Bartram "the greatest Natural Botanist in the World." Fagin 2.

16. Fagin 10.

17. William Bartram, *Travels,* ed. Mark van Doren (New York: Dover, 1955) 19. Unless otherwise identified, citations to William Bartram's *Travels* refer to this edition. *Travels* presents few textual problems. I prefer the widely available Dover edition to the out-of-print Yale (see note 14).

18. Ewan, *Drawings* 20.

19. Ewan, *Drawings* 11.

20. Fagin 11.

21. John W. Harshberger, *The Botanists of Philadelphia and Their Work* (Philadelphia: n.p., 1899) 87.

22. Joseph Ewan, "Early History," *A Short History of Botany in*

the U.S., ed. Joseph Ewan (New York: Hafner, 1969) 26.

23. Fagin 25.

24. Ewan, *Drawings* 3–34, 11, 9.

25. Thomas S. Kuhn, *Structure of Scientific Revolutions* (Chicago: Univ. of Chicago Press, 1962) 1.

26. Berkeley and Berkeley 61–76.

27. Bartram, *Travels* 15.

28. Arthur O. Lovejoy, *The Great Chain of Being* (Cambridge, Mass.: Harvard Univ. Press, 1964) 183.

29. Bartram, *Travels* 15.

30. Bartram, *Travels* 15.

31. Bartram, *Travels* 16.

32. Peter Kalm, *Peter Kalm's Travels in North America: The English Version of 1770,* ed. Adolph B. Benson, 2 vols. (New York: Wilson-Erickson, 1937) 2: 642.

33. Bartram, *Travels* 18.

34. Ewan, *Drawings* plates 12 and 22.

35. Philip C. Ritterbush, *Overtures to Biology: The Speculations of Eighteenth-Century Naturalists* (New Haven and London: Yale Univ. Press, 1964) 1.

36. François Delaporte, *Nature's Second Kingdom: Explorations of Vegetality in the Eighteenth Century,* trans. Arthur Goldhammer (Cambridge, Mass: MIT Press, 1982) 5.

37. Ritterbush 59.

38. Bartram, *Travels* 21.

39. Bartram, *Travels* 24–25.

40. Bartram, *Travels* 26.

41. Berkeley and Berkeley 36.

42. Bartram, *Travels* 369–70.

43. Frans A. Stafleu, *Linnaeus and the Linnaeans* (Utrecht: International Association for Plant Taxonomy, 1971) 118.

44. Bartram, *Travels* 370, 369n.

45. William Bartram, *Travels: A Facsimile of the 1792 London Edition* (Savannah, Ga.: Beehive Press, 1973) plate 1, facing page 18. Illustration from William Bartram, *Travels,* ed. Mark van Doren (New York: Dover, 1955). Reprinted by permission of the publisher.

46. Michel Foucault, *The Order of Things* (New York: Vintage, 1973) 133.

47. Foucault 134.

48. Foucault 135.

49. Bartram, *Travels* 33–55.

50. Bartram, *Travels* 33–34.

51. Bartram, *Travels* 35.

52. William Bartram, "Travels in Georgia and Florida, 1773–74: A Report to Dr. John Fothergill," *Transactions of the American Philosophical Society* n.s. 33 (1943):135.

53. Bartram, *Travels* 35.

54. See, for example, L. Hugh Moore, the latest in a brief tradition of Bartram critics who finds that Bartram's science is less than, and a part of, his romantic theories, particularly the "eighteenth-century aesthetic theory of the beautiful, sublime, and picturesque." Moore is right in pointing to the importance of these aesthetic categories. He has mistaken how they relate to Bartram's science, and blurred over the varying endorsements that each of them gets in the *Travels*. "The Aesthetic Theory of William Bartram," *Essays in Arts and Sciences* 9 (1983): 18.

55. Janice G. Schimmelman, "A Checklist of European Treatises on Art and Essays on Aesthetics Available in America through 1815," *Proceedings of the American Antiquarian Society* 93 (1983): 116–21.

56. Bartram, *Travels* 29.

57. Bartram, *Travels* 30.

58. Edmund Burke, *A Philosophical Enquiry into the Origin of Our Ideas of the Sublime and Beautiful*, ed. J. T. Boulton (Notre Dame: Univ. of Notre Dame Press, 1968) 57–58.

59. Burke 57–86.

60. Bartram, *Travels* 36; Burke 31.

61. Burke 5.

62. S. H. Monk, *The Sublime: A Study of Critical Theories in 18th-Century England* (New York: MLA, 1935) 42.

63. Burke 54.

64. Burke 11, 13, 15.

65. Bartram, *Travels* 30, 56, 64, 65, 73, 94, 102, 107, 131, 160, 163, 196, 279, 286, 274, 316, 341, 363.

66. Monk 92.

67. Christopher Hussey, *The Picturesque* (New York: Putnam, 1927) 1.

68. Hussey 83.

69. Bartram, *Travels* 150.

70. Bartram, *Travels* 151.
71. Bartram, *Travels* 155–56.
72. Bartram, *Travels* 94.
73. My discussion of Bartram's treatment of the native Americans is indebted to Mary Louise Pratt's identification and discussion of the manners-and-customs portrait in her "Scratches on the Face of the Country; or, What Mr. Barrow Saw in the Land of the Bushmen," *Critical Inquiry* 12 (1985): 120.
74. Bartram, *Travels* 44.
75. Bartram, *Travels* 45.
76. Roy Harvey Pearce, *The Savages of America* (Baltimore: Johns Hopkins Press, 1965) 143. Pearce assumes that all of Bartram's descriptions depict the natives as noble savages. He sees no distinction between the accounts obviously influenced by literary models and those influenced by natural historical models.
77. Bartram, *Travels* 201.
78. "The World was all before them, where to choose/Their place of rest, and Providence their guide," *Paradise Lost* XII. 646–47.
79. Bartram, *Travels* 289.
80. Bartram, *Travels* 290.
81. Bartram, *Travels* 110, 111, 113.
82. Bartram, *Travels* 164–65.
83. Bartram, *Travels* 164.
84. Bartram, *Travels* 164–65.
85. Bartram, *Travels* 250–51, 269, 296.
86. Bartram, *Travels* 301–2.
87. Bartram, *Travels* 357–62.
88. Bartram, *Travels* 366–67.
89. Bartram, *Travels* 284.
90. Bartram, *Travels* 285–86.
91. Bartram, *Travels* facing 381; separate title page of part 4.
92. William Bartram, "Observations on the Creek and Cherokee Indians," *Transactions of the American Ethnological Society* 3 (1853): 1–81.
93. Bartram, *Travels* 15.
94. Bartram, *Travels* 198.
95. Ewan, *Drawings* 36.

CHAPTER THREE. JEFFERSON AND THE DEPARTMENT OF MAN

1. Dumas Malone, *Jefferson the Virginian* (Boston: Little, Brown, 1948) 290.

2. William Peden, Introduction, *Notes on the State of Virginia*, by Thomas Jefferson (New York: W. W. Norton, 1972) xii. All citations to Jefferson's *Notes* refer to this edition.

3. Peden xvi.

4. Jefferson, *Notes* xx–xxi. Jefferson never published his projected revision.

5. John C. Greene, *American Science in the Age of Jefferson* (Ames: Iowa State Univ. Press, 1984) 33.

6. Edwin T. Martin, *Thomas Jefferson: Scientist* (New York: Henry Schuman, 1952).

7. Jefferson's description of the agrarian ideal is, for Leo Marx, "a point of view for which an accepted literary convention [the pastoral] is available." More generally Marx finds in Jefferson's *Notes* a "syntax of the middle landscape," a "perfect expression of the American pastoral ethos." *The Machine in the Garden* (New York: Oxford Univ. Press, 1964) 126, 121. Several critics concentrate on the two most anthologized passages in the work—the descriptions of the Natural Bridge and the Blue Ridge Gap. Floyd Ogburn, Jr. analyzes them using the linguistic tools of foregrounding and collocation. He discovers "the deep structure of pastoral" in these passages, where "all words collocate" around a single "nodal item"—nature. "Structure and Meaning in Thomas Jefferson's *Notes on the State of Virginia*," *Early American Literature* 15 (1980):144. Taking the two passages as emblematic, Harold Hellenbrand claims that *Notes* "embodies the mixed intentions of . . . complex pastoral, . . . the doubleness of Jefferson's attitude toward phenomenal nature and its stable core." He makes little effort to attach this attitude to Jefferson's scientific practice. "Roads to Happiness: Rhetorical and Philosophical Design in Jefferson's *Notes on the State of Virginia*," *Early American Literature* 20 (1985):19.

8. Marx apologizes for the book's "dense, dry, fact-laden surface" (118). Ogburn acknowledges the natural historical style in *Notes*, then deliberately ignores it when he analyzes what he calls the text's "stylistically aberrant passages" (142). Hellenbrand admits Jefferson's "dedication to facts" but claims that it is merely "apparent" (20).

9. Jefferson, *Notes* 143.

10. Merrill D. Peterson, *Thomas Jefferson and the New Nation* (New York: Oxford Univ. Press, 1970) 5.

11. Peden xviii.

12. Garry Wills, *Inventing America: Jefferson's Declaration of Independence* (New York: Doubleday, 1978) 167.

13. Malone 51.

14. Peterson 12.

15. Peterson 149.

16. Malone 439.

17. Edwin Morris Betts, ed., *Thomas Jefferson's Farm Book* (Philadelphia: American Philosophical Society, 1953) 324, 29.

18. Ann Leighton. *American Gardens in the Eighteenth Century* (Amherst: Univ. of Massachusetts Press, 1986) 210.

19. Betts, *Jefferson's Farm Book* 148.

20. Edwin Morris Betts, ed., *Thomas Jefferson's Garden Book* (Philadelphia: American Philosophical Society, 1985) 109, 282, 418, 450, 461, 502, 528, 527, 619, 380.

21. Betts, *Jefferson's Garden Book* 109, 155, 188, 172.

22. Betts, *Jefferson's Farm Book* 510.

23. Greene 197.

24. Quoted in Brooke Hindle, *The Pursuit of Science in Revolutionary America 1735–1799* (Chapel Hill: Univ. of North Carolina Press, 1956) 319.

25. Betts, *Jefferson's Garden Book* 528.

26. Richard Beale Davis, *Literature and Society in Early Virginia* (Baton Rouge: Louisiana State Univ. Press, 1973) 203.

27. He omitted a third faculty, imagination, and its human expression in the fine arts.

28. Robert A. Ferguson, *Law and Letters in American Culture* (Cambridge, Mass.: Harvard Univ. Press, 1984). Citing as models the legal works of Pufendorf, Burlamaqui, Montesquieu, and Blackstone, Ferguson traces the consequences of "Jefferson's decision to use the general form of these writings to provide a structure and organization to *Notes.*" He argues that this incorporation of the structural and organizational elements of the law places Jefferson at the beginning of a long line of American lawyer-writers: Trumbull, Tyler, Brackenridge, Brockden Brown, Irving, Bryant, Webster, and the Richard Henry Danas.

29. Ferguson 41.

30. Ferguson 40.

31. Ferguson 41.

32. Ferguson 48. Emphasis added.

33. Ferguson 47.

34. Myra Jehlen has explored the eighteenth-century postulate that the vigor of any civilization founded in a given climate depends on that climate. *American Incarnation* (Cambridge, Mass.: Harvard Univ. Press, 1986) 49–57.

35. Jefferson, *Notes* 279–80 n. 4.

36. Ferguson 46.

37. Ferguson 53.

38. Gilbert Chinard, ed., *The Commonplace Book of Thomas Jefferson* (Baltimore, Md.: Johns Hopkins Press, 1926) no. 47: 71– 72.

39. Betts, *Jefferson's Garden Book* 528.

40. Julian P. Boyd, ed., *The Papers of Thomas Jefferson* (Princeton, Princeton Univ. Press, 1950–) 8: 16.

41. Jefferson, *Notes* 39.

42. Jefferson, *Notes* 26ff, 38, 50–52, 66–69, 74, 83, 84, 86, 89, 103–7, 94–95, 108–9, 119, 144–45, 167, 171, 173, 179–96.

43. Jefferson, *Notes* 49.

44. Michel Foucault, *The Order of Things* (New York: Vintage, 1973) 134.

45. Garry Wills, *Inventing America: Jefferson's Declaration of Independence* (New York: Doubleday, 1978) 93–95.

46. Jefferson, *Notes* 65.

47. Jefferson, *Notes* 53–54.

48. Jefferson, *Notes* 50.

49. Jefferson, *Notes* 33, 80–81, 22.

50. Jefferson, *Notes* 20.

51. Jefferson, *Notes* 80, 102

52. Betts, *Jefferson's Garden Book* 528.

53. Betts, *Jefferson's Farm Book* 239.

54. Jefferson, *Notes* 143.

55. Jefferson, *Notes* 143.

56. Thomas Philbrick, "Thomas Jefferson" *American Literature 1764–1789: The Revolutionary Years,* ed. Everett Emerson (Madison: Univ. of Wisconsin Press, 1977) 166.

57. Stephen Jay Gould, *The Mismeasure of Man* (New York: W. W. Norton, 1981), 335–36.

58. Jefferson, *Notes* 138–42.

59. Jefferson, *Notes* 138.
60. Peterson 262.
61. Boyd 13: 651.
62. Gunnar Broberg, "Linnaeus's Classification of Man," *Linnaeus: The Man and His Work*, ed. Tore Frängsmyr (Berkeley: Univ. of California Press, 1983) 163, 167.
63. Jefferson, *Notes* 140–1.
64. Jefferson, *Notes* 163.
65. Jefferson, *Notes* 162.
66. Davis 203.
67. Jefferson, *Notes* 142.
68. Jefferson, *Notes* 87.
69. Jefferson, *Notes,* Query VIII, 280–81 n. 3.
70. Jefferson, *Notes* 163.
71. Malone 60.
72. Peterson 11.
73. Malone 285.
74. Jefferson, *Notes* 92–107.
75. Jefferson, *Notes* 94–95, 103–7.
76. Jefferson, *Notes* 102.
77. Jefferson, *Notes* 97.
78. Jefferson, *Notes* 101–2.
79. Stanley R. Hauer, "Thomas Jefferson and the Anglo-Saxon Language," *PMLA* 98 (1983): 891.
80. Thomas Jefferson, "Essay on the Anglo-Saxon Language," *The Writings of Thomas Jefferson,* ed. Albert Ellery Bergh, 20 vols. (Washington, D.C.: Thomas Jefferson Memorial Association, 1904) 18: 365–66.
81. Peterson 586.
82. Peterson 587.
83. Jefferson, *Notes* 62. Italics substituted for quotation marks.
84. Jefferson, *Notes* 62.
85. Malone 386.
86. Jehlen remarks on the "extraordinary" nature of this claim. It is proof of Jefferson's willingness to view the Other as endowed equally with whites. *American Incarnation* 48.
87. Jefferson, *Notes* 63.
88. Malone 385–86.
89. Peterson 247–48.
90. Jefferson, *Notes* xxi. For a recent interpretation of this quo-

tation that reemphasizes the "provisional" nature of *Notes,* see Robert Lawson-Peebles, *Landscape and Written Expression in Revolutionary America* (Cambridge: Cambridge Univ. Press, 1988) 183.

91. Peterson 239, 242.

92. Jefferson, *Notes* 62.

CHAPTER FOUR. CRÈVECOEUR'S "CURIOUS OBSERVATIONS OF THE NATURALIST"

1. Thomas Philbrick, *St. John de Crèvecoeur* (New York: Twayne, 1970) 42.

2. D. H. Lawrence, *Studies in Classic American Literature* (1923; New York: Viking, 1964) 23–33.

3. Julian P. Boyd, ed., *The Papers of Thomas Jefferson* (Princeton: Princeton Univ. Press, 1954–) 8: 155–56.

4. J. Hector St. John de Crèvecoeur, *Letters from an American Farmer and Sketches of 18th-Century America,* ed. Albert E. Stone (New York: Penguin, 1963) 39. Subsequent citations are to this edition.

5. Jean F. Beranger suggests that Mr. F. B. resembles Dr. Fothergill, patron of John Bartram. "The Desire of Communication: Narrator and Narratee in *Letters from an American Farmer,*" *Early American Literature* 12 (1977): 74.

6. Gay Wilson Allen and Roger Asselineau, *St. John de Crèvecoeur: The Life of an American Farmer* (New York: Viking, 1987) 5, 6–7, 10–11, 21.

7. Allen and Asselineau 13.

8. Allen and Asselineau 26.

9. Allen and Asselineau 30–31.

10. Allen and Asselineau 34–35, 59, 62.

11. Bernard Chevignard, "St. John de Crèvecoeur in the Looking Glass: *Letters from an American Farmer* and the Making of a Man of Letters," *Early American Literature* 19 (1984): 175.

12. Allen and Asselineau 62. For another view, see Chevignard (186–87), who claims that after he arrived in France Crèvecoeur significantly revised the *Letters* from the notes and journals he hid in the bottom of the boxes.

13. Albert E. Stone, Introduction, *Letters from an American Farmer and Sketches of 18th-Century America,* by J. Hector St. John de Crèvecoeur (New York: Penguin, 1981) 12.

14. Quoted by Allen and Asselineau 211.

15. Quoted by Chevignard 175.

16. Chevignard 177.

17. William Martin Smallwood, *Natural History and the American Mind* (New York: Columbia Univ. Press, 1941) 91–92.

18. Allen and Asselineau 77, 78, 89–91, 133, 196.

19. Boyd 8: 155.

20. Allen and Asselineau 157.

21. Chevignard 182.

22. Crèvecoeur 42.

23. Crèvecoeur 165.

24. Crèvecoeur 90.

25. Richard Beale Davis, *Literature and Society in Early Virginia* (Baton Rouge: Louisiana State Univ. Press, 1973) 203.

26. See, for example, Philbrick 80, for a typical description of life on the islands as "idyllic."

27. Philbrick 81.

28. Crèvecoeur 108.

29. Crèvecoeur 110.

30. Crèvecoeur 114.

31. Crèvecoeur 113–14.

32. Crèvecoeur 118.

33. Crèvecoeur 120–22, 125.

34. Crèvecoeur 160.

35. Crèvecoeur 154.

36. See, again, Philbrick 81.

37. Crèvecoeur 167, 174–75.

38. Arthur O. Lovejoy, *The Great Chain of Being* (Cambridge, Mass.: Harvard Univ. Press, 1964) 184–85. Lovejoy quotes Addison (*Spectator* 519), who calls plentitude "the Capacity of Being."

39. Crèvecoeur 177.

40. Crèvecoeur 53.

41. Crèvecoeur 180.

42. On 4 April 1794 the members of the American Philosophical Society heard Benjamin Smith Barton read a paper on New World snakes in which he reviewed the long history of their supposed ability to enchant their prey. "A Memoir Concerning the Fascinating Faculty which has been ascribed to the Rattle-Snake, and other American Serpents," *Transactions of the American Philosophical Society* 4 (1799): 74–113.

43. Crèvecoeur 184.

44. Lawrence 29; Elayne Antler Rapping, "Theory and Experience in Crèvecoeur's America," *American Quarterly* 19 (1967): 712; Philbrick 83.

45. Crèvecoeur 181, 184.

46. Crèvecoeur 178.

47. Rapping 713; Philbrick 63; Annette Kolodny, *The Lay of the Land* (Chapel Hill: Univ. of North Carolina Press, 1975) 60–61; David Robinson, "Crèvecoeur's James: The Education of an American Farmer," *Journal of English and Germanic Philology* 80 (1981): 562–63.

48. Crèvecoeur 187–88.

49. Crèvecoeur 191.

50. Crèvecoeur 194.

51. Myra Jehlen, "J. Hector St. John Crèvecoeur: A Monarcho-Anarchist in Revolutionary America," *American Quarterly* 31 (1979): 210, 218.

52. Crèvecoeur 203.

53. Crèvecoeur 204.

54. Crèvecoeur 218–21.

55. Philbrick 87.

56. For instance, Joel R. Kehler, "Crèvecoeur's Farmer James: A Reappraisal," *Essays in Literature* 3 (1976): 212.

57. Crèvecoeur 211, 219, 225.

58. Crèvecoeur 222.

59. Crèvecoeur 215.

60. Crèvecoeur 213.

61. Crèvecoeur 226.

62. Steven J. Dick, *Plurality of Worlds* (Cambridge: Cambridge Univ. Press, 1982) 142–75.

63. Crèvecoeur 227.

64. Crèvecoeur 225–26.

65. Crèvecoeur 69.

66. Crèvecoeur 67.

67. Crèvecoeur 70, 71.

68. Crèvecoeur 72, 77.

69. Crèvecoeur 79, 82, 90, 71.

70. Crèvecoeur 76.

71. Lawrence 30.

72. Allen and Asselineau 89.

73. Allen and Asselineau 229.

74. Stone 7.

75. Robert P. Winston cites Albert Stone, Jr., and Harry B. Henderson III, then himself examines *Letters* as a romance. "'Strange Order of Things!': The Journey to Chaos in *Letters from an American Farmer*," *Early American Literature* 13 (1984–85): 249.

76. Henry Mackenzie, *The Man of Feeling* (London: Oxford Univ. Press, 1970) 3.

CHAPTER FIVE. THE PASSING OF NATURAL HISTORY AND THE LITERATURE OF PLACE

1. John Walton, *John Filson of Kentucke* (Lexington: Univ. of Kentucky Press, 1956) 21, 24–25, 29, 117.

2. John Filson, *The Discovery and Settlement of Kentucke* (Ann Arbor: University Microfilms, 1966) 113–18.

3. Filson 27.

4. Filson 23–24, 28.

5. Filson 11.

6. Filson 52.

7. Filson 58–78.

8. Filson 64–65, 62.

9. Filson 70, 73.

10. The white speaker at the Pinkasaw Council (pp. 82–84) assumes the native Americans' style of oratory.

11. Filson 81.

12. Filson 87–109.

13. Willard Rouse Jillson, ed., *The Boone Narrative: The Story of the Origin and Discovery Coupled with the Reproduction in Facsimile of a Rare Item of Early Kentuckiana; to which is Appended a Sketch of Boone and a Bibliography of 238 Titles* (Louisville, Ky.: Standard Printing Company, 1932).

14. Richard Slotkin traces the growth of the Boone myth in his *Regeneration through Violence: The Mythology of the American Frontier 1600–1860* (Middletown, Conn.: Wesleyan Univ. Press, 1973) 268–312.

15. Filson 109.

16. Filson 36.

17. Michel Foucault, *The Order of Things* (New York: Vintage, 1970) 137–38.

18. Edwin Morris Betts, ed., *Thomas Jefferson's Garden Book, 1766–1824* (Philadelphia: American Philosophical Society, 1944) 529.

19. Foucault 269.

20. Foucault 267, 268.

21. Foucault 269.

22. The Canadian government now calls "Eskimos," which has come to be perceived as a derogatory term, "Inuits," which is what they call themselves.

23. Brooke Hindle, "Charles Willson Peale's Science and Technology," *Charles Willson Peale and His World* (New York: Harry N. Abrams, 1982) 113.

24. Lillian B. Miller, "Charles Willson Peale: A Life of Harmony and Purpose," *Charles Willson Peale and His World* (New York: Harry N. Abrams, 1982) 193.

25. Dillon Ripley, *The Sacred Grove* (New York: Simon and Schuster, 1969) 35.

26. William Martin Smallwood, *Natural History and the American Mind* (New York: Columbia Univ. Press, 1941) 108.

27. Hindle 129–34.

28. Stephen A. Tyler, "Post-Modern Ethnography: From Document of the Occult to Occult Document, " *Writing Culture,* eds. James Clifford and George E. Marcus (Berkeley: Univ. of California Press, 1986) 130.

29. Clifford Geertz, *Works and Lives: The Anthropologist as Author* (Stanford: Stanford Univ. Press, 1988) 135.

30. Geertz 132.

31. Geertz 140.

32. Henry David Thoreau, *Walden,* ed. J. Lyndon Shanley (Princeton: Princeton Univ. Press, 1971) 4.

33. Thoreau, *Walden* 68.

34. William Bartram, *Travels*, ed. Mark van Doren (New York: Dover, 1955) 221–49, for example.

35. Thoreau, *Walden* 225.

36. *Essay on Man* I.264.

37. Thoreau, *Walden* 226.

38. Thoreau, *Walden* 234.

39. Henry David Thoreau, *Winter,* ed. H.G.O. Blake (Cambridge, Mass.: 1887) 405–6.

40. Explorations of this tradition can be found in Robert Finch and John Elder, eds., *The Norton Book of Nature Writing* (New York: W. W. Norton, 1990); and in Peter A. Fritzell, *Nature Writing and America* (Ames: Iowa State Univ. Press, 1990).

Works Cited

Adair, James. *Adair's History of the American Indians*. New York: Promontory Press, 1930.

Allen, Gay Wilson and Roger Asselineau. *St. John de Crèvecoeur: The Life of an American Farmer.* New York: Viking, 1987.

Arner, Robert D. "Pastoral Patterns in Bartram's *Travels*." *Tennessee Studies in Literature* 18 (1973): 133–46.

Barnett, Louise K. *The Ignoble Savage: American Literary Racism, 1790–1890.* Westport, Conn.: Greenwood Press, 1975.

Barnett, Thomas. "William Bartram and the Age of Sensibility." Diss. Georgia State Univ., 1982.

Barnhart, John Henry. "Bartram Bibliography." *Bartonia* 12 (1931): 51–67.

Barton, Benjamin Smith. "A Memoir Concerning the Fascinating Faculty which has been ascribed to the Rattle-Snake, and other American Serpents." *Transactions of the American Philosophical Society* 4 (1799): 74–113.

Bartram, John. "Diary of a Journey through the Carolinas, Georgia, and Florida from July 1, 1765 to April 10, 1766." *Transactions of the American Philosophical Society* n.s. 33 (1942): 1–120.

Bartram, William. "Observations on the Creek and Cherokee Indians." *Transactions of the American Ethnological Society* 3 (1853): 1–81.

———. *Travels*. Ed. Gordon DeWolf. Savannah, Ga: Beehive, 1973.

———. *Travels*. Ed. Mark Van Doren. New York: Dover, 1955.

———. "Travels in Georgia and Florida, 1773–74: A Report to Dr. John Fothergill." *Transactions of the American Philosophical Society* n.s. 33 (1943): 121–242.

———. *Travels: Naturalist's Edition*. Ed. Francis Harper. New Haven: Yale, 1958.

Batten, Charles L. *Pleasurable Instruction: Form and Convention in Eighteenth-Century Travel Literature*. Berkeley: Univ. of California Press, 1978.

Beranger, Jean F. "The Desire of Communication: Narrator and Narratee in *Letters from an American Farmer.*" *Early American Literature* 12 (1977) 73–85.

Berkeley, Edmund and Dorothy Smith Berkeley. *The Life and Travels of John Bartram.* Tallahassee: Univ. Presses of Florida, 1982.

Betts, Edwin Morris, ed. *Thomas Jefferson's Farm Book.* Philadelphia: American Philosophical Society, 1953.

———. *Thomas Jefferson's Garden Book, 1766–1824.* Philadelphia: American Philosophical Society, 1985.

Bewley, Marius. *The Eccentric Design.* New York: Columbia Univ. Press, 1959.

Boyd, Julian P., ed. *The Papers of Thomas Jefferson.* 20 vols. Princeton: Princeton Univ. Press, 1950–.

Broberg, Gunnar. "Homo sapiens: Linnaeus's Classification of Man." *Linnaeus: The Man and His Work.* Ed. Tore Frängsmyr. Berkeley: Univ. of California Press, 1983. 156–94.

Burke, Edmund. *A Philosophical Enquiry into the Origin of Our Ideas of the Sublime and the Beautiful.* Ed. J. T. Boulton. Notre Dame: Univ. of Notre Dame Press, 1968.

Carver, Jonathan. *Travels through the Interior Parts of North America in the Years 1766, 1767, and 1768.* 3rd ed. London: n.p., 1781.

Chevignard, Bernard. "St John de Crèvecoeur in the Looking Glass: *Letters from an American Farmer* and the Making of a Man of Letters." *Early American Literature* 19 (1984): 173–90.

Chinard, Golbert, ed. *The Commonplace Book of Thomas Jefferson.* Baltimore: John[s] Hopkins Press, 1926.

Cooper, Lane. "A Glance at Wordsworth's Reading." *Modern Language Notes* 22 (April 1907): 110–17.

Crèvecoeur, J. Hector St. John de. *Letters from an American Farmer and Sketches of 18th-Century America.* Ed. Albert E. Stone. New York: Penguin, 1963.

Darlington, William. *Memorials of John Bartram and Humphrey Marshall.* Philadelphia: Lindsay and Blackiston, 1849.

Davis, Richard Beale. *Literature and Society in Early Virginia.* Baton Rouge: Louisiana State Univ. Press, 1973.

Delaporte, François. *Nature's Second Kingdom: Explorations of Vegetality in the Eighteenth Century.* Trans. Arthur Goldhammer. Cambridge, Mass.: MIT Press, 1982.

Dick, Steven J. *Plurality of Worlds: The Extraterrestrial Life Debate from Democritus to Kant.* Cambridge: Cambridge Univ. Press, 1982.

Earnest, Ernest. *John and William Bartram: Botanists and Explorers.*

Philadelphia: Univ. of Pennsylvania Press, 1940.
Ewan, Joseph. "Early History." *A Short History of Botany in the United States*. Ed. Joseph Ewan. New York: Hafner, 1969. 27–48.
———, ed. *William Bartram: Botanical and Zoological Drawings, 1756–1788*. Philadelphia: American Philosophical Society, 1968.
Fagin, N. Bryllion. *William Bartram: Interpreter of the American Landscape*. Baltimore: The Johns Hopkins Press, 1933.
Ferguson, Robert. *Law and Letters in American Culture*. Cambridge, Mass.: Harvard Univ. Press, 1984.
Filson, John. *The Discovery and Settlement of Kentucke*. Ann Arbor: University Microfilms, 1966.
Finch, Robert and John Elder, eds. *The Norton Book of Nature Writing*. New York: W. W. Norton, 1990.
Forster, John Reinhold. "Account of Several Quadrupeds from Hudson's Bay." *Selected Works by Eighteenth-Century Naturalists and Travellers*. Ed. Kier B. Sterling. New York: Arno, 1974. Unpaginated.
Foucault, Michel. *The Order of Things*. Trans. Michel Foucault. New York: Vintage, 1973.
Franklin, Wayne. *Discoverers, Explorers, Settlers: The Diligent Writers of Early America*. Chicago and London: Univ. of Chicago Press, 1979.
Fritzell, Peter A. *Nature Writing and America*. Ames: Iowa State Univ. Press, 1990.
Geertz, Clifford. *Works and Lives: The Anthropologist as Author.* Stanford: Stanford Univ. Press, 1988.
Greene, John C. *American Science in the Age of Jefferson*. Ames: Iowa State Univ. Press, 1984.
Gordon, Captain Harry. "Journal of Captain Harry Gordon." *Travel in the American Colonies, 1600–1783*. Ed. Newton D. Mereness. New York: Macmillan, 1916. 455–89.
Gould, Stephen Jay. *The Mismeasure of Man*. New York: W. W. Norton, 1981.
Harshberger, John W. *The Botanists of Philadelphia and Their Work*. Philadelphia: n. p., 1899.
Hauer, Stanley R. "Thomas Jefferson and the Anglo-Saxon Language." *PMLA* 98 (1983): 879–98.
Hellenbrand, Harold. "Roads to Happiness: Rhetorical and Philosophical Design in Jefferson's *Notes on the State of Virginia*." *Early American Literature* 20 (1985): 3–23.
Hindle, Brooke. "Charles Willson Peale's Science and Technology."

Charles Willson Peale and His World. New York: Harry N. Abrams, 1982. 106–69.

———. *The Pursuit of Science in Revolutionary America, 1735–1789.* Chapel Hill: Univ. of North Carolina Press, 1956.

Hussey, Christopher. *The Picturesque.* New York: Putnam, 1927.

Jefferson, Thomas. "Essay on the Anglo-Saxon Language." *The Writings of Thomas Jefferson.* 20 vols. Ed. Albert Ellery Bergh. Washington, D.C.: Thomas Jefferson Memorial Association, 1903–5. 18: 359–411.

———. *Notes on the State of Virginia.* Ed. William Peden. New York: W. W. Norton, 1972.

Jehlen, Myra. *American Incarnation.* Cambrige, Mass.: Harvard Univ. Press., 1986.

———. "J. Hector St. John Crèvecoeur: A Monarcho-Anarchist in Revolutionary America." *American Quarterly* 31 (1979) 204–22.

Jillson, Willard Rouse, ed. *The Boone Narrative: The Story of the Origin and Discovery Coupled with the Reproduction in Facsimile of a Rare Item of Early Kentuckiana; to which is Appended a Sketch of Boone and a Bibliography of 238 Titles.* Louisville, Ky: Standard Printing Company, 1932.

Kalm, Peter. *Peter Kalm's Travels in North America: The English Version of 1770.* Ed. Adolph B. Benson. 2 vols. New York: Wilson-Erickson, 1937.

Kehler, Joel R. "Crèvecoeur's Farmer James: A Reappraisal." *Essays in Literature* 3 (1976): 206–12.

Kolodny, Annette. *The Lay of the Land.* Chapel Hill: Univ. of North Carolina Press, 1975.

Kuhn, Thomas S. *The Structure of Scientific Revolutions.* Chicago: Univ. of Chicago Press, 1962.

Lawrence, D. H. *Studies in Classic American Literature.* 1923. New York: Viking, 1964.

Lawson-Peebles, Robert. *Landscape and Written Expression in Revolutionary America.* Cambridge: Cambridge Univ. Press, 1988.

Leighton, Ann. *American Gardens in the Eighteenth Century.* Amherst: Univ. of Massachusetts Press, 1986.

Lindroth, Sten. "The Two Faces of Linnaeus." *Linnaeus: The Man and His Work.* Ed. Tore Frängsmyr. Berkeley: Univ. of California Press, 1983.

Linné, Carl von. *Species Plantarum: A Facsimile of the First Edition 1753.* 2 vols. London: The Ray Society, 1957.

Looby, Christopher. "The Constitution of Nature: Taxonomy as Politics in Jefferson, Peale, and Bartram." *Early American Literature* 22 (1987): 252–73.

Lovejoy, Arthur O. *The Great Chain of Being*. Cambridge, Mass.: Harvard Univ. Press, 1973.

———. "The Place of Linnaeus in the History of Science." *Popular Science Monthly* 71 (1907): 498–508.

Lowes, John Livingston. *The Road to Xanadu*. Boston and New York: Houghton Mifflin Company, 1930.

Mackenzie, Henry. *The Man of Feeling*. 1771. London: Oxford Unv. Press, 1970.

Malone, Dumas. *Jefferson the Virginian*. Boston: Little, Brown, 1948.

Martin, Edwin T. *Thomas Jefferson: Scientist*. New York: Henry Schuman, 1952.

Marx, Leo. *The Machine in the Garden: Technology and the Pastoral Ideal in America*. New York: Oxford Univ. Press, 1964.

Medeiros, Patricia M. "Three Travelers: Carver, Bartram, and Woolman." *American Literature 1764–1789: The Revolutionary Years*. Ed. Everett Emerson. Madison: Univ. of Wisconsin Press, 1977. 195–212.

Mereness, Newton D. *Travels in the American Colonies*. New York: Macmillan, 1916.

Miller, Lillian B. "Charles Willson Peale: A Life of Harmony and Purpose." *Charles Willson Peale and His World*. New York: Harry N. Abrams, 1982; 170–233.

Monk, S. H. *The Sublime: A Study of Critical Theories in 18th-Century England*. New York: MLA, 1935.

Moore, L. Hugh. "The Aesthetic Theory of William Bartram." *Essays in Arts and Sciences* 9 (1983): 17–35.

Nye, Russel Blaine. *The Cultural Life of the New Nation 1776–1830*. New York: Harper and Brothers, 1960.

Ogburn, Floyd. "Structure and Meaning in Thomas Jefferson's *Notes on the State of Virginia*." *Early American Literature* 15 (1980): 140–50.

Pearce, Roy Harvey. *The Savages of America: A Study of the Indian and the Idea of Civilization*. Baltimore: Johns Hopkins Press, 1965.

Peden, William. Introduction. *Notes on the State of Virginia*. By Thomas Jefferson. New York: W. W. Norton, 1982.

Peterson, Merrill D. *Thomas Jefferson and the New Nation*. New York: Oxford Univ. Press, 1970.

Philbrick, Thomas. *St. John de Crèvecoeur*. New York: Twayne, 1970.

———. "Thomas Jefferson." *American Literature 1764–1789: The Revolutionary Years*. Ed. Everett Emerson. Madison: Univ. of Wisconsin Press, 1977, 145–70.

Pratt, Mary Louise. "Scratches on the Face of the Country; or, What Mr. Barrow Saw in the Land of the Bushmen." *Critical Inquiry* 12 (Autumn 1985): 119–43.

Rapping, Elayne Antler. "Theory and Experience in Crèvecoeur's America." *American Quarterly* 19 (1967): 705–18.

Reichel, Bishop John Frederick. "Travel Diary of Bishop and Mrs. Reichel and their Company." *Travel in the American Colonies 1690–1783*. Ed. Newton D. Mereness. New York: Macmillan, 1916, 583–99.

Ripley, Dillon. *The Sacred Grove*. New York: Simon and Schuster, 1969.

Ritterbush, Philip C. *Overtures to Biology: The Speculations of Eighteenth-century Naturalists*. New Haven and London: Yale Univ. Press, 1964.

Robinson, David. "Crèvecoeur's James: The Education of an American Farmer." *Journal of English and Germanic Philology* 80 (1981): 552–70.

Rogers, Robert. *A Concise Account of North America*. 1765. New York: Johnson Reprint Corporation, 1966.

———. *Journals of Robert Rogers*. 1765. N.p.: Readex Microprint, 1966.

Schimmelman, Janice C. "A Checklist of European Treatises on Art and Essays on Aesthetics Available in America through 1815." *Proceedings of the American Antiquarian Society* 93(1983): 134–45.

Slotkin, Richard. *Regeneration through Violence: The Mythology of the American Frontier, 1600–1860*. Middletown, Conn.: Wesleyan Univ. Press, 1973.

Smallwood, William Martin. *Natural History and the American Mind*. New York: Columbia Univ. Press, 1941.

Smith, William. *Historical Account of Bouquet's Expedition against the Ohio Indians in 1764*. Cincinnati: Robert Clarke Company, 1907.

Stafleu, Frans A. *Linnaeus and the Linnaeans*. Utrecht: International Association for Plant Taxonomy, 1971.

Stearn, William Thomas. Notes on the Illustrations. *Species Plantarum: A Facsimile of the First Edition 1753*. By Carl Linnaeus. 2 vols. London: The Ray Society, 1957, 2: 61–72.

———. "Sources, Format, Method and Language of the *Species Plantarum*." *Species Plantarum: A Facsimile of the First Edition 1753*. By Carl Linnaeus. 2 vols. London: The Ray Society, 1957. 1:81–96.

Sterling, Keir B., ed. *Contributions to the History of American Natural History*. New York: Arno, 1974.

———. *Selected Works by Eighteenth-Century Naturalists and Travellers*. New York: Arno, 1974.

Stilgoe, John R. "Fair Fields and Blasted Rock: American Land Classification Systems and Landscape Aesthetics." *American Studies* 22 (1981): 21–33.

Stone, Albert E. Introduction. *Letters from an American Farmer and Sketches of 18th-Century America*. By J. Hector St. John de Crèvecoeur. New York: Penguin, 1963.

Stork, William. *Account of East Florida*. London: n.p., 1766.

Thoreau, Henry David. *Walden*. Ed. J. Lyndon Shanley. Princeton: Princeton Univ. Press, 1971.

———. *Winter*. Ed. H. G. O. Blake. Cambridge, Mass.: Houghton, Mifflin and Company, 1887.

Tyler, Moses Coit. *The Literary History of the American Revolution 1763–1783*. 2 vols. New York: G. P. Putnam, 1898.

Tyler, Stephen A. "Post-Modern Ethnography: From Document of the Occult to Occult Document."' *Writing Culture*. Eds. James Clifford and George E. Marcus. Berkeley: Univ. of California Press, 1986, 122–40.

Walton, John. *John Filson of Kentucke*. Lexington: Univ. of Kentucky Press, 1956.

Wills, Garry. *Inventing America: Jefferson's Declaration of Independence*. New York: Doubleday, 1978.

Winston, Robert P. "'Strange order of things!' The Journey to Chaos in *Letters from an American Farmer*." *Early American Literature* 19 (1984–85): 249–67.

Index